Ur

Su

Modeling and Problem Solving Techniques for Engineers

Modeling and Problem Solving Techniques for Engineers

by

László Horváth and Imre J. Rudas

ELSEVIER
ACADEMIC
PRESS

Amsterdam   Boston   Heidelberg   London   New York   Oxford
Paris   San Diego   San Francisco   Singapore   Sydney   Tokyo

Elsevier Academic Press
200 Wheeler Road, 6th Floor, Burlington, MA 01803, USA
525 B Street, Suite 1900, San Diego, California 92101-4495, USA
84 Theobald's Road, London WC1X 8RR, UK

This book is printed on acid-free paper. ∞

Library of Congress Cataloging-in-Publication Data

British Library Cataloguing in Publication Data
A catalogue record for this book is available from the British Library

ISBN: 0-12-602250-X

For all information on all Academic Press publications
visit our Web site at www.academicpress.com

PRINTED IN THE UNITED STATES OF AMERICA

04   05   06   07   08   09   9 8 7 6 5 4 3 2 1

Table of Contents _____

CHAPTER **8**

Construction and Relating Solid Part Models in CAD/CAM Systems

CHAPTER **9**

Creating Kinematic Models in CAD/CAM Systems

Preface

Engineering design is more than an activity of skilled humans in industry: it is also a cultural mission in both common and technical senses. In traditional engineering, abundant time was available for engineers to merge their activity into national and global cultural environments. By the twentieth century the impact of technology and product development accelerated engineering activities. Technology became the prevailing aspect of engineering and the cultural aspect was overshadowed. In the meantime, despite the continued acceleration of product development, the technology-stimulated development of computering has given a great chance for the reintegration of the technical and cultural aspects of engineering by the automation of routine activities. Recently, customer demand driven engineering has forced engineers into considering both the cultural and social aspects of engineering.

A new style of engineering has been established, where advanced information and computer technologies are applied to handle product related engineering information in computer systems. Engineering activities are done virtually to the greatest extent possible. A virtual technology has emerged, primarily for design, analysis, manufacturing, and human–computer interaction purposes. The integrated and coordinated handling of information serves engineering activities from the first idea of a product to the last demand for product related information. Engineering modeling has become one of the activities that has a substantial effect on the achievements of company objectives such as minimal engineering costs, a short product development cycle, effective handling of the minimal number of product changes, reduced time to introduce new products, minimal cost of developing new products, improved quality of products, and advancements in competitiveness.

Several characteristics of the present style of engineering mirror recent enhancements of computer based work of engineers. Digital definition and simulation of products and manufacturing processes, and management of product lifecycle information are the main issues in model based engineering. Modeling is supported by CAD/CAM systems. A CAD/CAM system is configured for a well-defined segment of engineering. Advanced computer tools and engineering processes represent one of its typical functionalities. Mechanical parts are defined by application oriented form features with exact boundary-representation of the solid geometry. Local demands for the application of modeling are fulfilled by custom development of open architecture CAD/CAM systems in third party or company environments where domain and application specific knowledge is available. Proven methods are included for capturing, sharing, and reusing corporate knowledge throughout the entire engineering process. Knowledge embedded in models allows less experienced engineers to solve complex problems in product design, analysis, manufacturing planning, and production related engineering activities. Knowledge based modeling enables companies to capture and deploy their best practices through custom-built applications. This reduces the risk of product failure, production inefficiency, market mismatches, and after-delivery compliance costs. Alternative modeling processes and product variants can be handled. By the use of realistic simulation, companies can anticipate behaviors of their future products and manufacturing process operations, evaluate multiple design concepts, and create design variants from a common concept. Better understanding of product, production, and market by the application of advanced analysis and simulation, better and more human culture oriented product design, better utilization of earlier design concepts, and fast response to customer demands are also important effects. The creating, handling, and control of engineering information in modeling systems help to coordinate managing activities in several departments, in a well-organized concurrent engineering system where engineering activities are done simultaneously and any activity can start immediately when input information is made available by other engineering activities. Now, all sizes of businesses are covered. Advanced modeling is not a privilege of powerful international companies: it is also accessible to small and medium-sized businesses, especially in global collaboration. Integrated modeling resources and models offer a powerful background for industrial engineering.

In the 1970s and 1980s, huge numbers of excellent computer methods and programs were conceptualized as potential tools to make advancements in actual problem solving for engineering. Most of them remained a dream because of the available computer performance and application technology. By the 1990s, advances in computing performance, computer graphics, software technology, and communication in computer systems produced the necessary computer technology for turning earlier dreams into reality. The result was a rapid take-off of industrially applied virtual technology at the end of the twentieth century. Modeling and model based simulation, together with intelligent computing, facilitate handling of the characteristics and behavior of modeled objects so that advanced models as virtual prototypes replace physical prototypes at evaluations of product design.

This book is intended to help practicing engineers to get a high level start in engineering modeling. It is also intended to serve as a textbook for undergraduate and graduate students. It serves master students at the great step between bachelor and master courses. The authors were motivated to assist engineers and students who are not computer application specialists in understanding modeling. Inspired by virtual engineering technology, new ideas are created by a new generation of engineers. This is why students must be prepared for modeling. Advanced computer methods and tools are also required for experienced engineers to enhance problem solving. Unfortunately, most experienced engineers cannot understand modeling to the extent necessary for its efficient application. They cannot utilize the inherent design power of virtual technology in their work. Consequently, they work in one of the conventional ways or their skills and experience are not utilized in engineering activities. This choice of bad and worse can be avoided at companies if experienced engineers are better assisted in understanding the virtual world of advanced computers. Both engineers and employers are motivated by a great chance for improved personal career and business.

Numerous excellent and popular books have been written about engineering modeling for beginners and interested people in the past two decades. At the other end of the offerings, excellent books are available for people who are specialists in the development of modeling procedures and computer aided engineering systems. There is a wide gap between these two important purposes in the recent and presently available choice of books in computer aided engineering. Experienced engineers fail in hunting for books that would help them in understanding modeling and in its practical application for their real world industrial engineering tasks. High level books are not needed with deep explanations of ideas and methods for engineers who just would like to understand and utilize modeling methods and tools and who would not like to be specialists in the theoretical and practical development of modeling and related systems. This textbook is intended to be a contribution to fill the gap.

One of the main emphases in this book is on knowledge assisted group work of engineers in integrated, application area oriented, and worldwide networked engineering modeling systems. This style of computer aided engineering utilizes the main results of the development efforts in engineering applications of computer systems during the 1990s. Readers of this book will find text directed toward users of advanced modeling offered by existing, widely applied, well-accepted, and leading industrial CAD/CAM systems. At the same time, this book does not intend to compete with but refers to excellent books on general CAD/CAM concepts, and on deep mathematical analysis of geometrical models and intelligent computing approaches. Because engineering modeling is so dynamic, the content of this book focuses on up-to-date modeling and cannot devote too much text to the historical pioneering modeling methods from the 1970s and 1980s that are not necessary to understand present modeling.

Introduction

Comprehensive application software systems serve virtual engineering activities in industrial practice. These systems are called traditionally CAD/CAM (computer aided design/computer aided machining (or manufacture)). They apply advanced computer modeling. The name CAD/CAM refers to their primary application in the past: geometric model based planning of computer controlled machining. CAD/CAM has developed from several isolated or interfaced computer aided engineering activities to integrated virtual engineering in local, wide area, or global computer systems. Anyway, the primary purpose of virtual engineering is high level and effective assistance for engineering decisions in an increasingly competitive environment. Engineering design tasks can be considered as a chain of engineering decisions. An engineering decision requires information about the complexity of engineering objects, about the effects of earlier decisions on the actual decision, and about the effects of the actual decision on the affected engineering objects. Continuous product development to withstand competition has changed the style of engineering decision making in recent years. Frequent revision of decisions is required by changed or new customer demand, new developments, and changed company strategy. Decisions often reveal the need for revision of some earlier decisions. Engineering decisions need assistance from powerful computer modeling enhanced by intelligent computing where sophisticated model descriptions of objects and their relationships and knowledge based reasoning are available. The growing importance and wide application of virtual engineering need a deep knowledge of modeling by engineers practicing in industry and applying modeling or changing it in the future.

Advanced CAD/CAM systems support the coexistence of computer based modeling and engineering design. Not so long ago the prevailing medium for the communication of engineers was engineering drawing in its copied form of the blueprint. By the 1950s and 1960s engineering tasks became increasingly complex in their concepts, details, and considerations. Engineers had won battles in ideas and innovations; however, they had lost battles in information and information processing. Moreover, engineers were under constant pressure to make innovation cycles shorter and shorter. At the same time, ideas and innovative solutions demanded more and more data be described and processed. In the early 1950s the application of modeling became unavoidable. Intricate shapes emerged on parts to be machined by automatically controlled machine tools. They could not be described by conventional engineering drawings and blueprints. The only way was mathematical description of the shape and application of this information for the calculation of cutting tool paths. This is considered as the starting point of virtual engineering. Shape and tool path modeling were followed by modeling of other engineering objects as complete parts, assemblies, kinematics, finite elements, etc. Models of mechanical parts have moved into applications far beyond tool path generation. Now, a huge number of analysis tools rely on sophisticated models of analyzed engineering objects, their environments, and simulation processes. The real world of engineering is simulated in an advanced and comprehensive virtual engineering environment. Models as advanced descriptions of engineering objects represent much more than new media that replace blueprints and computer generated drawings. The fantastic development of communication in the past decade has had a strong effect on virtual engineering. Models are exchanged between CAD/CAM systems in a worldwide integrated engineering environment. At the same time, centralized databases provide services for a lot of engineering workstations in different geographical sites. Remote modeling systems are operated by humans using special browser surfaces opened to the Internet. The utmost purpose of engineering modeling is the product model that covers all the information demanded by all possible engineering activities during the product lifecycle from the first sketch by the designer to recycling.

This book covers concepts, approaches, principles, and practical methods for an area of integrated engineering activities where mechanical parts and their structures are conceptualized, designed, analyzed, and planned and programmed for computer controlled manufacturing. Application of the introduced virtual technology is not restricted to traditional mechanical engineering but includes all industries where sophisticated mechanical parts are applied in products, such as cars, household appliances, home electronics, sports equipment, furniture, watches, toys, etc. The design, analysis, and manufacturing of shapes and methods for the production of such items are central because they still constitute the majority of engineering related computer applications at industrial companies. Because the covered area is so large, special emphasis is placed on principles and methods of outstanding industrial significance anticipated for the next few years. When topics and issues were selected, advanced and proven CAD/CAM systems from industrial practice were considered.

Chapter 1 presents a quick glance at the essential concepts of the magic world of computer aided engineering. Chapter 2 introduces computer model based engineering activities with a basic package of related knowledge. Chapters 3–6 explain the basic understanding of models for the description of engineering objects. Chapters 7–9 give detailed discussion about computing methods for the creation of model entities that have importance in the present and future practice of modeling, analysis and manufacturing planning.

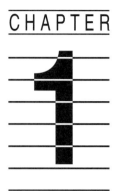

The Magic World of Virtual Engineering

An engineer is sitting in front of a screen, points to symbols and lines, and types some data. The computer shows a mechanical part in three dimensions allowing the engineer to develop or modify its model or to include it in an assembly structure. All of these activities use the same set of simple interactive functions. Seeing this style of engineering design, one might consider the work of an engineer as easy but not very exciting. Others seeing the result may consider the engineer as a magician or think of engineering as a simple handling of magic design automata. The truth is that human knowledge, skill, and experience are utilized in an interactive communication with a computer system that is capable of creating a model description by utilizing the knowledge, skill, and experience of the engineers who created the computer and the modeling system.

As stated in the Introduction, the application of computers by engineers started with the challenging task of the design, manufacturing planning, and computer controlled manufacturing of intricate surfaces, mainly for aerodynamic purposes. The first widely applied computer assistance for engineering served drawing, calculations, data storage, and documentation printing activities during

1

the 1960s and 1970s. The next milestone was wide acceptance of computer models in the 1980s. The modeling tools[1] in industrially applied CAD/CAM systems of the 1970s and 1980s executed direct commands from engineers, mainly for creating line, surface, etc., of geometric model entities. In the 1990s, engineering oriented shape modeling by parametric application features was introduced. Modeling methods also featured in integrated assembly, analysis, and manufacturing planning. In recent years, intelligent features have been given to advanced modeling. Object oriented and knowledge supported modeling tools, among others, analyze the behavior of modeled objects, propose alternative solutions, navigate interactive model creation, and prevent obviously erroneous results. To do this, sophisticated, integrated, knowledge based,[2] and comprehensive sets of modeling tools and models are available for engineers. A new scene of engineering has emerged in computer systems. It is called the virtual world[3] and it gives a new and enhanced quality to engineering.

1.1   The Virtual World for Virtual Engineering

The virtual world is featured by *product and process centered modeling* in contrast to the task orientation of earlier modeling. Although product models are capable of representing any product related information, they are configured according to the demands of their applications. Despite their potential complexity, product models can range from the description of several modeled objects to the representation of very complex and interrelated structures of various engineering objects. Advanced models act as *virtual*

---

[1]The phrase "modeling tool" is applied to a procedure that serves a well-defined modeling purpose.

[2]Knowledge based modeling utilizes modeling tools that are capable of doing some engineering activities automatically, using built-in knowledge.

[3]The word "virtual" refers to a computer resident world.

prototypes. Sophisticated modeling and analysis techniques make it possible to move a great deal of prototype development from expensive machine shops into virtual worlds. Engineers are still the most important participants of virtual worlds. They conduct orchestras of hundreds of modeling procedures through effective graphical human–computer interface actions (HCIs).

Engineering objects in the virtual world are described by using methods from virtual technologies such as computer aided modeling, intelligent computing, simulation, data management, multimedia assisted computer graphics, and human–computer interaction. The behavior of engineering objects is assessed under various real-world situations. The dynamic nature of the development of virtual technologies is the result of their applications in dynamically developing areas of business. Questions concerning the real content of the virtual world for industrial practice will be answered in the subsequent chapters of this text.

The *virtual world* for engineering (Figure 1-1) is constituted of interrelated descriptions of engineering objects as parts, assemblies, kinematics, analysis results, manufacturing processes, production

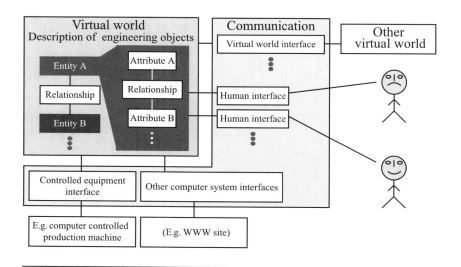

Figure 1-1   Virtual world.

equipment, manufacturing tools, instruments, etc. Elements of the descriptions are entities and their attributes. The structure of the descriptions is defined as the relationships between entities, or their attributes. Operated by humans interacting with computer procedures, the virtual world communicates with controlled equipment, other computer systems, and other virtual worlds. Production by automatic equipment is controlled by direct application of information from the virtual world. An implementation of a virtual world is considered as a special computer system.

The following is a simple sequence of activities for creating a virtual world as an example from everyday modeling.

A part is conceptualized as a set of shape objects.

A shape object is described as a form feature entity. It is related to geometric model entities in its representation.

Relationships are defined between form features and their dimension attributes.

A human controls part model creation procedures by interactive communication.

The model is communicated with a World Wide Web (WWW) site where it is visualized.

The purpose of the part model is computer controlled manufacturing of a computer numerical controlled (CNC) machine tool.

Geometric model entities and their relationships are illustrated by the example of Figure 1-2. A simple assembly consists of *Part 1* and *Part 2*. During the creation of a model of *Part 2*, surface entities FS_1 and FS_2 are related by their common boundary line CE_1. Parts are related using relationships referring to geometric entities. The contact relationship RC_1 defines contact between surfaces FS_2 and FS_3. A set of geometric entities is available in each modeling system.

Construction of a virtual world in the course of engineering activities starts from ideas about the system to be modeled and the engineering objects to be included in it. Typically, objects and their structures are defined. The basic approach to construction may be

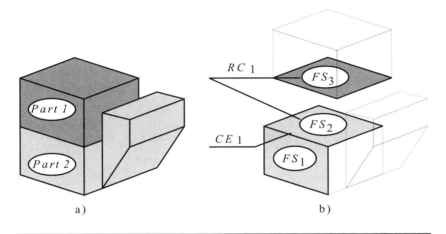

a)   b)

Figure 1-2   Relating entities.

top-down, bottom-up, or mixed. Following a pure top-down approach, modeling starts with the definition of a structure, and then objects are created for elements of the structure. Following a pure bottom-up approach, first objects are created and then their structure is defined. A mixed approach represents everyday practice where some objects are available at the start, then structure is defined, and finally the remaining objects are created according to the structure. Examples for predefined elements can be units of computers, integrated circuits, fasteners, bearings, and other standard elements of mechanical and other engineering systems. Typical structures can be predefined, stored, retrieved, and adapted for individual tasks.

This text discusses the virtual world where the main characteristics of dominating objects are shape related and other objects can be characterized in connection with shape-centered objects. Because real-world shapes are three-dimensional, their computer descriptions should be three-dimensional (3D) for the virtual world. Shapes are described in a space called model space. In Figure 1-3, an object is positioned in a model space. Non-shape features of an object such as stress (ST) and temperature (TE) at a given point of its volume and properties of a surface in

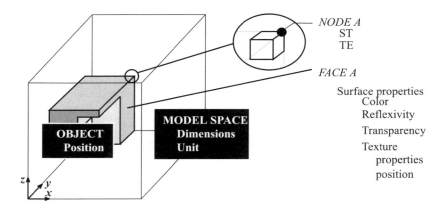

Figure 1-3   Shape based description of engineering objects.

its boundary are mapped to appropriate shape model entities as attributes.

One of the exceptions to shape based engineering objects is an electronic system where active and passive circuit elements are connected by routes but not in a dimensioned space. However, this model of an electronic system is completed with shape models of printed circuit board arrangements and programming of automatic assembly and inspection equipment. At automatic assembly, circuit elements *R31*, *R33*, and *C33* should be positioned relative to the printed circuit board (Figure 1-4). At automatic inspection, a camera image of the ready assembled printed circuit board is compared with a master image to reveal parts omitted during assembly.

Figure 1-5 outlines the description of a mechanical part in a *virtual world*. A virtual space with dimensions of the real space accommodates objects. The object in this example is a part, one of the product related engineering objects. Shape, position, material, surface properties, outside relationships, and behavior of the object are described. Outside relationships include connections to other parts such as contact of surfaces, possibilities of movements relative to other parts such as linear movement along a slide, and restraints such as a pin. Behaviors of a part are defined by stress,

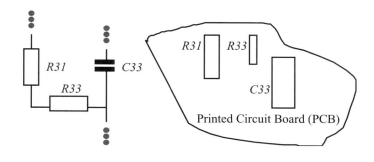

Figure 1-4   Modeling of electronic circuits.

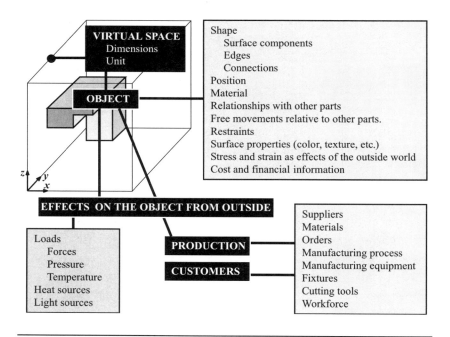

Figure 1-5   Description of an object in the virtual world.

strain, temperature, cost, etc., at different effects of the outside world on the object such as loads, etc. The effects of the outside world are modeled to know the behavior of the object. In addition, the virtual world includes production system objects and their relations to product objects. Finally, customers are in connection with production through marketing, demand forecast, and sales services. Humans are involved in the design, manufacturing, and application of products.

Engineering is only one but perhaps the largest application area of virtual worlds. Other important application areas are medical treatments, flight simulators, and highway traffic control systems, for example. Entertainment applications include live shows, movie and television productions, and theater performances. In addition, virtual arts are emerging.

1.2 Let Us Go into Virtual

Competition forces industry to make fast and frequent changes of product design. We can say that the prototype development stage has expanded to the entire life of the product. Times allowed by the market for product changes are too short for expensive and time-consuming physical product prototyping. Creating new or altered physical prototypes is impossible within the available time frame: throughput times and the related manufacturing costs are unrealistic. Conventional physical prototyping cannot cope with the present requirements for product development. The only solution is to move prototyping into the virtual world to the greatest extent. In recent years, companies have converted most of their conventional design and prototyping activities for advanced products into virtual technology. Analyses of engineering objects using their correct and sophisticated models are robust enough to replace most physical prototyping. Moreover, the virtual prototype describes much more information about a product than a conventional physical prototype. However, a suitable virtual environment is also expensive, so cost and value analyses are needed to assist decisions about prototyping technology. Analyses of technical and financial reasons for virtual and physical prototyping generally propose the application of a mixture of them. Changes of the human mind, imagination, and fantasy as a result of creativity as well as fast responses to changes in market, finance, or production conditions change the product through integrated virtual engineering activities. The results and experiences from physical prototyping are accumulated in virtual prototyping systems and applied to enhance virtual prototyping and to decrease the demand for physical prototyping of similar engineering tasks.

It is a standard capability of recent industrial engineering systems to create products from ideas in an automatic, but human governed way. Highly automated engineering and production require integration of all the actual engineering activities with the related manufacturing process. Integration assumes modeling procedures that communicate a mutual understanding of input and output information, and use the same database and the same user interface (Figure 1-6).

Two virtual related concepts in manufacturing are *virtual manufacturing* and *virtual controlled manufacturing*. Virtual manufacturing is an assessment of the product and its manufacturing process without any physical manufacturing or measurements. Virtual controlled manufacturing is the control of real-world manufacturing equipment by use of information from the virtual world. Figure 1-7 gives a simplified explanation of this scenario. Humans communicate their intents and concepts with a computer system. Procedures translate these concepts into the virtual world, analyze the models, translate the models into control programs and control production equipment. The virtual prototype[4] is created and manufactured virtually by appropriate analysis of the manufacturability of modeled objects. Finally, virtually controlled manufacturing is applied to create physical prototypes of modeled objects using computer controlled production equipment.

---

[4]The virtual prototype is sometimes cited as the virtual product.

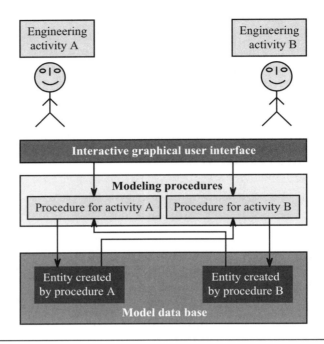

Figure 1-6 Integration of engineering activities.

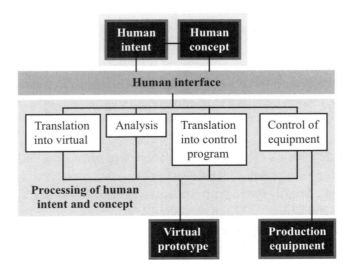

Figure 1-7 Manufacturing and the virtual world.

Concepts of visual reality and virtual reality are sometimes confused. Virtual reality is a sophisticated computer description of real-world conditions in the virtual world. It cannot communicate directly with humans because it has been developed for the purpose of communication between computer procedures in the form of data structures. Visual reality is the tool that converts these data structures into graphic or other understandable forms in order to visualize computer descriptions for humans. Concepts and intents originate in humans in visual form: visual reality is the tool to translate them into a form understandable by computer procedures. Two-way interactive graphics-based communication is applied. Virtual and visual realities are key techniques for the representation of engineering objects and communication of represented information between humans and procedures in virtual worlds.

Figure 1-8 shows the human–computer–human communication chain during engineering modeling. The visual concepts of engineers are translated into model data sets and vice versa. This is why visual computing is one of the main areas of development in computer technology. Data oriented virtual worlds are visualized by use of impressive computer graphic tools. Engineers

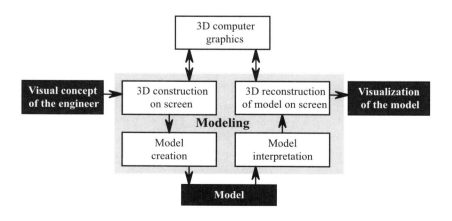

Figure 1-8   Visual reality assisted virtual reality.

communicate the model-creating procedures by the definition of 3D objects within a construction area on the screen called a view port. Model application procedures visualize the model for humans by 3D reconstruction of modeled objects in the view port.

Engineering modeling is highly shape and position intensive. The shape of an engineering object is described mathematically in a model space characterized by a coordinate system (Figure 1-9). Dimension driven modeling was developed during the 1980s for the purpose of shape definition by type and dimensions as well as by dimension definitions on existing shapes. Dimensions may control shapes. Shapes are positioned and oriented in the model space. The position and orientation of a modeled object in the space are defined absolutely or in relation to other modeled objects. Figure 1-9 shows three basic position definitions. The position can be defined in space by coordinates of a characteristic point such as *P1* of the

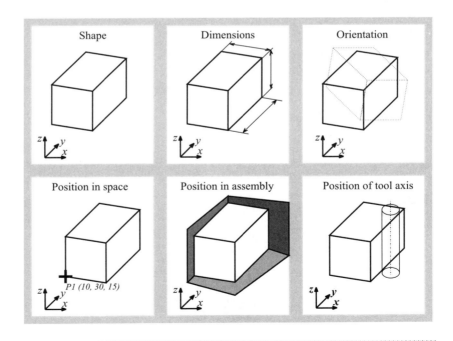

Figure 1-9   Shape and position information in models.

modeled object. One of the available position definitions for placing a part in an assembly uses three pairs of planes. Three planes of the other modeled object are in contact with the appropriate planes of the current modeled object. The third positioning is of the cutting tool in relation to the actual machined surface of the modeled part.

1.3   A Great Change in Engineering Activities

The introduction of electronic drawing based computer aided engineering signified a great change in engineering activities during the 1960s. No more drawings, blueprints, stacks of books, or finding solutions from old drawings and documents were necessary. No more tedious and time-consuming manual drawing with many repetitions of the same drawing elements, such as screws, was needed. The introduction of modeling technologies was the next dramatic change in engineering. No more drawings are necessary on paper sheets or screens that accept crazy designs without any reaction. Instead, there is a magic screen with a human governed virtual world behind it. Regiments of computer procedures are ready to fulfill the wishes of engineers. However, wishes must be suitable for processing by the appropriate procedures. Procedures check and sometimes do not accept human decisions on the basis of the intent of the engineers who developed and configured the procedures. An engineer who is not able to understand the computer procedures or cannot interpret actual problems with those procedures is helpless. Engineers often prefer less advanced procedures because they allow creation of models of simpler and even incorrect engineering objects. As a consequence, engineers sometimes are susceptible to accepting less advanced computer procedures and blaming ones that are more advanced but hard to understand for less experienced engineers. However, a product intended to compete successfully must be advanced under any circumstances. Similarly to other magic worlds, the virtual world does not work without magicians.

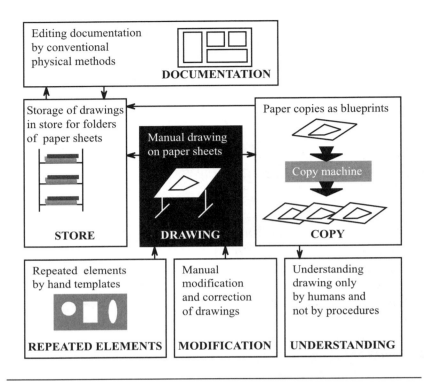

Figure 1-10 Manual drawing.

The capabilities of the traditional means of putting ideas on paper manually, on drawing boards using the communication tool of engineering drawing, are summarized in Figure 1-10. Drawings are stored in the form of paper sheets. Archives are hard to handle and demand expensive storage. Documents are copied from paper to paper, traditionally in the form of blueprints. Templates of drawing elements such as circles, ellipses, rectangles, contours of a screw or nut, etc., can be repeated many times. Modifications for the correction or improvement of drawings are done by physical erasing, overwriting, etc. Documents are edited manually or by conventional printing technology. Modified or similar parts need separate drawings for elaboration. The only purpose of drawings is that of communication between humans.

The change to drawing on computer by the use of drawing software resulted in a tremendous saving of human work. Drawings are edited on screen by interactive graphics and all modifications are as simple as redefining drawing elements (Figure 1-11). Repeated elements, drawing details, and complete drawings are created, combined, stored, and reused electronically. Plotters and printers produce hard copies. Drawings are visualized on screen. The handling of large and complicated drawings is made easier by zoom and pan functions. Hyperlinks can be defined to any other documents accessible on the Internet. Drawing files can be placed in folders of file systems of operating systems or in databases. Special editing programs produce relevant documentation.

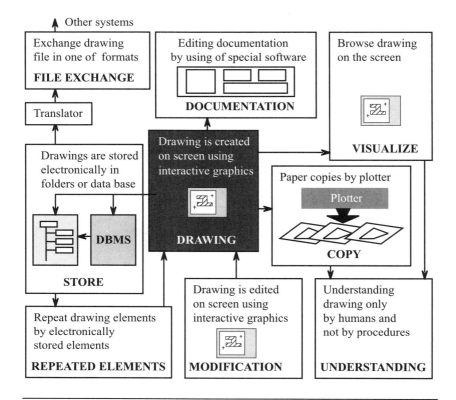

Figure 1-11   Drawing based computer aided engineering.

Electronic documentation utilizes multimedia. Drawing files are exchanged between different drawing systems in one of the standard formats, or converted between formats of different drawing systems. Translators are available for data conversion between different drawing file formats as building elements of CAD/CAM systems. Electronic drawing utilizes the conventional symbols of engineering drawing so that it is an excellent communication medium between humans. It is not suitable for communication between modeling procedures, although simple contour information can be extracted for additional processing. Making models of parts, for example, can use contours extracted from electronic drawings.

When computer based drawing methods were implemented at industrial companies, paper based drawings were captured in computer systems and converted to electronic form (Figure 1-12). This was tedious work so creating new drawings often proved the better solution. Electronic drawings and their hard copies are created

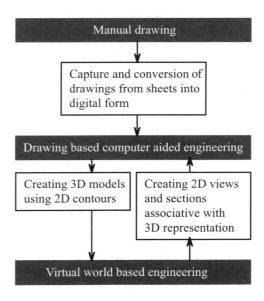

Figure 1-12   Conversion between drawings and models.

primarily for people outside of computer systems, for example those in conventional manufacturing, by manually operated machine tools and manual mechanical and electronic assembly.

Let us take the first step towards becoming a magician by understanding the world behind the screen of interactive modeling (Figure 1-13). A human develops an idea about the forthcoming object to be modeled, selects appropriate model entity creation procedures, gives values for parameters of procedures, then the procedure creates and combines model entities. Parameters in this context are inputs of entity creation procedures and control operation of these procedures according to the actual model creation task and human intent. Navigators offer suitable values of actual parameters for selection. Model entity data sets are stored in the model data pool temporarily, during the modeling session, and then they are stored in the database as results of the modeling session. Abandoned variants, sketches, etc., are discarded or saved in a separate place for later retrieval and use in other, similar tasks.

Model files are exchanged between modeling systems in the original format, in the format of another system, or in one of the standard neutral formats. Except for the first case, translation (conversion) of model data between different formats is necessary. Translation into a neutral format is followed by a second translation into the format understandable by the receiving system. Models are created to facilitate communication between modeling procedures. Humans understand the model during its creation and modification through the interactive graphics user interface. Advanced visualization associated with easy to use prototype modeling procedures is often called a digital mock-up. Traditionally, realistic visualization by interactive graphics user interfaces was not a primary objective because of its relatively high demand for computing performance. In recent years, the application of powerful graphical processors in computer systems has brought a demand for enhanced quality of visualization. Anyway, modeling procedures display on the screen only the information necessary for the actual human interaction. Too much, irrelevant, or badly organized information makes the screen chaotic.

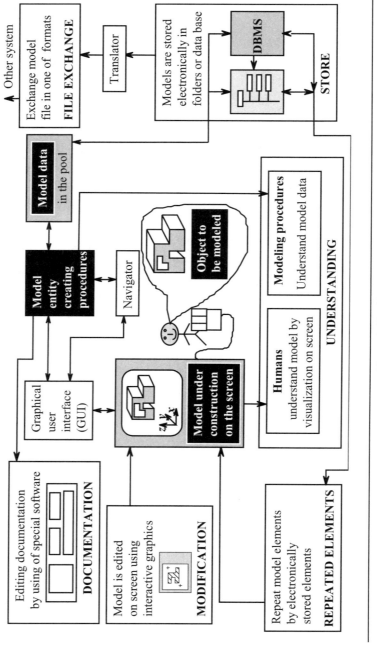

Figure 1-13   Modeling.

1.4   The Model Space

Model descriptions of shape based engineering objects are created in model spaces. A model space, also called a pool (Figure 1-14), is a dedicated temporary data storage that is active during modeling sessions. The engineer constructs a model using real dimensions of the modeled objects. Any position within the model space is defined by the x, y, z *model coordinate system*, which is a Cartesian type global coordinate system. The model coordinate system is sometimes called a *world coordinate system* referring to the world of modeling which is defined by the actual model space. For comfortable position and dimension definition, engineers define *local coordinate systems* on modeled objects or at other appropriate positions in the model space. In Figure 1-14, a local coordinate system (x_L, y_L, z_L) is applied for easy construction of the part model starting from the upper face of the part. Coordinates defined in a local coordinate system are automatically transformed into coordinates in the model coordinate system.

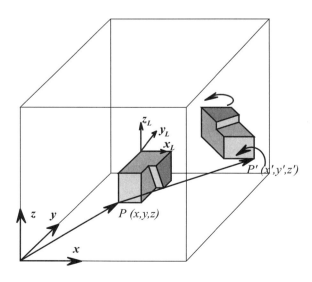

Figure 1-14   Model space.

Objects are reoriented and repositioned in the model space using simple transformations such as translation, rotation, and their combinations. In Figure 1-14 point P is defined by x, y, and z global coordinates. Following this, P is translated to point P' and the object is rotated in two steps around the y and z axes of the model coordinate system. If a series of transformations of an object is time programmed, the object is animated.

Engineers control modeling procedures. They need visualization of any shape related information of the model under construction, at any time. Communication and interaction between humans and computer procedures (HCI) use a graphical user interface (GUI). Significant computer resources are devoted to interactive vector graphics that visualize shape model representations by graphical entities. 3D objects are projected from the model space to an appropriately positioned two-dimensional (2D) screen area. Elementary graphical entities are lines (vectors), polylines (open and closed chains of vectors), and filled-in surface areas (Figure 1-15). Advanced graphics handle higher-level geometric entities such as curves, surfaces, and solids. During modeling sessions, visualization serves interactions. It concentrates on edges and other characteristic lines of modeled objects (Figure 1-15a). Contour lines of the selected and actual objects are highlighted. Hidden edges often must be visible to allow selection by a pointing device, as in Figure 1-15a. When hidden edges disturb the clear picture of the object, they can be made invisible (Figure 1-15b). When realistic visualization of an object is needed, its filled surfaces are made visible. Appearance of a surface depends on its color, material properties, and illumination. Illumination of objects in the model space is provided by light source model entities. Figure 1-15c shows uniform color and light intensity on all points of each surface. To achieve a more realistic visualization, the surface must be divided into small areas to show changes of light intensity and color along the surface.

The viewpoint of an engineer is often changed during model creation sessions depending on the actual region of the object under work or analysis. In Figure 1-15d, the viewpoint has been

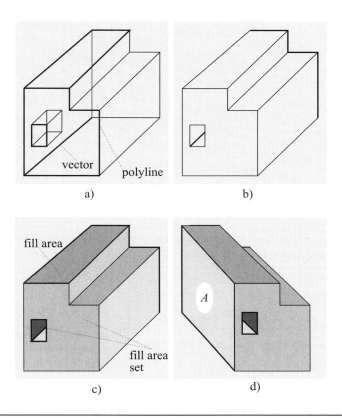

Figure 1-15 Visualization of the model space.

changed to bring surface *A* to the foreground. A pointing device such as a mouse is used to grip and rotate objects on the screen. This action changes the actual viewpoint of the engineer, but does not change the position and orientation of the object in the model space because the model space rotates with the object.

CHAPTER

2

Activities in Virtual Engineering

The purpose of this chapter is to give an insight into engineering modeling activities by step-by-step explanation, in accordance with emphases in present industrial practice. Basic concepts are explained to prepare readers for Chapters 3–9 that discuss areas of modeling in detail.

Computer modeling intensive engineering is applied in a new industrial environment where configurable variants of products are designed, analyzed, manufacturing planned, and production planned and manufactured using leading information and computer technology. Products have changed from heavy and relatively simple mechanical and electric structures to complex mechatronics where styled shapes are accompanied by well-engineered functional elements. Modeling tools are utilized in the continuous, market-demanded improvement of product and production related capabilities.

One of the possible groupings of engineering and related company activities is shown in Figure 2-1. The purpose of this grouping is to introduce virtual engineering by the main application areas at industrial companies. Group No. 1 involves engineering activities in

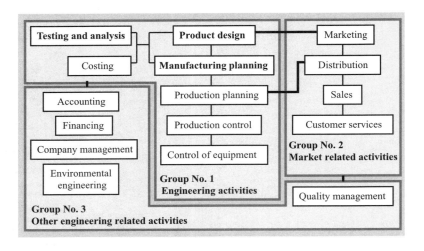

Figure 2-1   Engineering and related company activities.

the strict sense of the word. Design of the product is assisted by testing and analysis as well as costing. Product design is done concurrently with manufacturing planning. Manufacturing plans and orders are processed into production plans and schedules by production planning. Production is controlled by use of these production plans and schedules. Automatic production equipment is controlled by execution of control programs downloaded into machine shops according to production plans and schedules. Market related activities (Group No. 2) are marketing, distribution of products, sales activities, and customer services. Other engineering related activities (Group No. 3) are accounting, financing, environmental engineering, and company management. This book focuses on activities bold typed in the Group No. 1 (Figure 2-1).

2.1   Computer Representations for Shape-centered Engineering

Shape-centered design, analysis, and manufacturing planning rely on shape description of parts as engineering objects. Shapes are organized into structures. Non-shape information is mapped to these structures and the shape descriptions. The shape model is constructed in the model space by the definition of elementary shapes such as contours, elementary surfaces, solid primitives, and form features. Elementary shapes are constructed by a set of lines, curves, and surfaces organized by topological structure. Curves and surfaces are described by mathematical functions. Elementary shapes are combined into complex shapes of parts. Animation is applied to move the resultant shape in the model space according to a time-scheduled program. Contacts and other relationships are defined between parts to describe the relation of parts in assemblies. Finally, relative movements between parts are described by kinematics. The above outlined description of a mechanical system is completed by finite element modeling for the purpose of understanding the effects of loads and restraints on parts, such as stress, strain, and temperature. A mesh of finite elements can be created on the surface or in the volume of a part. Characteristics of part surfaces that affect appearance are also described in the model and are applied by computer graphics as realistic visualization of engineering objects.

Figure 2-2 gives a survey of shape-centered models of a mechanical system. Objects O_1, O_2, and O_3 are combined into complex shape O_4. O_4, O_5, and O_6 constitute an assembly. C_1 in O_2 is described as a parametric curve. Parts are in contact at their flat surfaces. O_5 is allowed to move relative to O_4 along curve C_2. Similarly, O_6 is allowed to move relative to O_4 along line L_1. Contacts of the flat surface pairs are not broken during allowed movements. A sequence of three positions of O_4 in the model space is described in time by Frames 1–3. Each frame is a discrete position of the animated object at a given time. The appearance of surfaces of O_6 is shown by shading.

The virtual world that is established by model based engineering is in close connection with the designed, stored, visualized, and documented worlds (Figure 2-3). The designed world is the physical world to be modeled. Modeling experts sometimes should remember that the final purpose of engineering activities is to

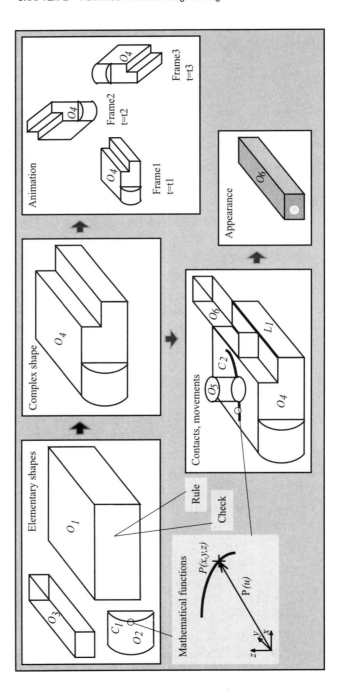

Figure 2-2   Basic model representations of shape-centered engineering objects.

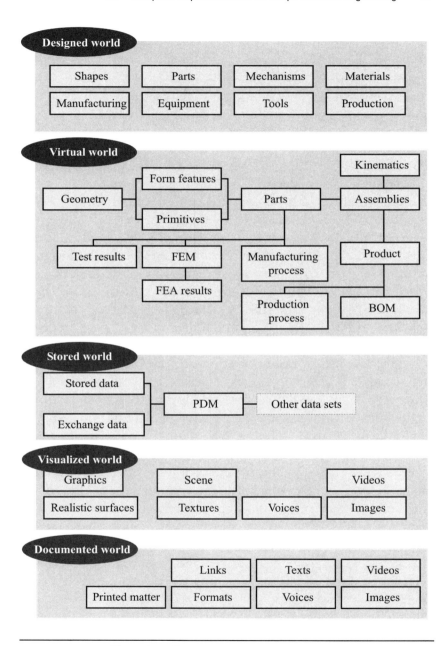

Figure 2-3   The scenario of model based design.

produce a physical world. Data sets generated during modeling are stored and handled for optimal retrieval and utilization of computer resources. One of the main differences between the application of the modeled and stored worlds is that engineers must understand the logic of modeling, but the logic of data handling is the competence of computer system managers. Product data management (PDM) focuses on product structures for variants and handling models in environments where models are created by several different modeling systems in the data organization for a product. The visualized world is for communications and interactions with engineers and applies efficient interactive graphics. Beware of a fantastic visualized world with a poor modeling background! Paper based and electronic documents are needed for both official purposes and the personal use of engineers. Visualization is in direct connection with modeling procedures and modeling is impossible without it while documents, electronic or traditional paper based, are used without any connection to modeling procedures.

The designed world includes product and production related engineering objects. Shapes, parts, mechanisms, and materials constitute products. The designed world also involves manufacturing and production processes as well as equipment and tools applied to the manufacturing of parts. The designed world is described in computer models. Design is product and production oriented while modeling is computer methodology oriented. In early model based design it was hard to harmonize the mathematical and computing oriented modeling with engineering oriented product design. This is why application orientation gained great emphasis in the development of modeling during the late 1980s and 1990s. The harmony of the designed and modeled worlds is a recent result of the development of model based engineering. Designs are represented in computers according to the communicated intent of design engineers. The designed worlds in the mind of the engineer and the modeled world behind the screen must be harmonized. One of the main purposes of this text is to assist this human related process. The main components of the virtual world are given below, as they are grouped and interconnected in Figure 2-3.

2.2 Models of Mechanical Units ⸻⸻⸻⸻⸻

The term mechanical unit relates to any units that contain parts placed in assemblies and, where applicable, allow movements between pairs of parts with a given degree of freedom. Mechanical units are required in all machines, instruments, cars, domestic appliances, home entertainment devices, etc. The modeling of mechanical units is applied in the engineering and production of most industrial products and production equipment as well as in devices for the manufacture of other products. For example, the same geometry is applied to describe curves in car bodies and in tools for making shoes.

Although several other geometric model representations prevailed in the earlier stages of the development of modeling, the boundary representation became the prevailing shape of modeling in engineering during the late 1980s and 1990s. The boundary geometric model representation uses topology and geometry for the description of shapes and consists of topological and geometrical entities.

On the right of Figure 2-4, the shape of a part is visualized. Any point and curve can be computed in the model coordinate system using the mathematical description of the shape. The shape in this example is covered by flat surfaces; it does not contain any curved surfaces or curves on its boundary. Lines and flat surfaces constitute a complete closed boundary of the body. The shape model is based on the principle of boundary representation with surfaces covering the shape and lines at the intersection of surfaces.

The analysis of the effect of the change of a geometric entity on other geometric entities needs information about the geometric entities adjoining it. In other words, the model must include information about connecting curves and surfaces. This information is carried by topological entities of the boundary representation.

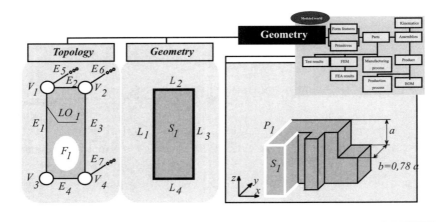

Figure 2-4   Boundary representation of the shape of a part.

Individually described point, curve, and surface geometrical entities are mapped to vertex, edge, and face topological entities,
respectively. In Figure 2-4, lines L_1–L_4, enclosing the surface S_1,
are mapped to edges E_1–E_4, respectively. Surface S_1 is mapped to
face F_1. Face F_1 is connected to the closed loop of edges LO_1. An
edge is connected to other edges by vertex topological entities at its
ends. A complete outside or inside boundary of a body is called a
shell. A shell is a closed and consistent structure of faces, edges,
and vertices. If material is defined inside a shell or between shells,
the resultant shape model is a solid representation. Some shapes
can be unambiguously defined by lines and curves mapped to
topological edges. This makes it possible to work with a reduced
set of topological and geometrical entities in wireframe models.
The reduced set includes point and curve (or line) geometric, and
vertex and edge topological entities.

Seeing the boundary of a body on the screen with surfaces and
their intersection lines, an engineer understands the shape easily.
This is not true for computer procedures because these can understand and process only predefined data placed in predefined structures. The consequence of any changes of a surface or line is the
change of several lines and surfaces in its neighborhood. Figure 2-5

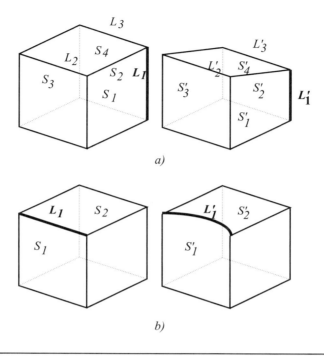

Figure 2-5   Effect of modification of a line mapped to a topological edge.

shows two different situations. In Figure 2-5a, change of length of the straight line L_1 to L_1' results in a change of the dimension or position of lines L_2 and L_3 to L_2' and L_3' and flat surfaces S_1–S_4 to S_1'–S_4', respectively. All geometric entities remain linear. Sometimes construction of the part starts with a simple outline shape then the final shape is created by a sequence of changes of linear entities to curved entities. In Figure 2-5b, the change of line L_1 to spatial curve L_1' changes flat surfaces S_1 and S_2 to curved surfaces S_1' and S_2', respectively.

Mechanical parts are defined by dimensions placed in the shape model too. Characteristic dimensions of shapes are used as parameters of entity creation procedures. Dimensions can be also defined on a shape, after its creation, between arbitrary points. Dimensional parameters of a shape or different shapes can be related in formulas. These formulas, such as $b = 0.78a$ in Figure 2-4, are placed in the shape model. The part size is changed by modification of one or more control parameters. Modification of one of the dimensions changes all related dimensions automatically, by use of the relationships of dimensions described in the formulas. Relationships between model entities or their attributes are called associativities. When an associativity, a dimension, or any other characteristic of any modeled object is fixed as a decision of the engineer, it can be defined as a constraint in the model. If relationships between dimensions are constrained, they are saved during model changes.

The next question is how to construct the shape model of a part during engineering design sessions. Two basic principles are applied in present engineering practice: combination of individual elementary shapes, and modification of an initial shape by form features. The prevailing method is shape modification; however, advanced hybrid modeling systems also offer element combination as an auxiliary method.

Using the method of element combination, the design engineer first defines elementary shapes for the desired functions of the part, then combines them into the final shape of the part. Elementary shapes are called primitives. Although the method of element combination can be applied to contours and surfaces, its main application area is solid modeling. On the upper part of Figure 2-6, two primitives are defined by their types and dimensions then fused into a complex shape. Primitives are created in their final position in the part under construction. When a primitive is created in a position other than its final one, an additional transformation should be included to its final position. Parameters of a primitive can be also defined by selection of lines on an existing shape. In Figure 2-6, the base circle C of the cylinder primitive is selected and transformed to the end of the edge line l of the box primitive as C'. C' serves as a parameter at the creation of a third cylinder primitive. Element combination with half space representation of primitives was the only solid modeling method during the 1970s and early 1980s. Element combination in recent modeling

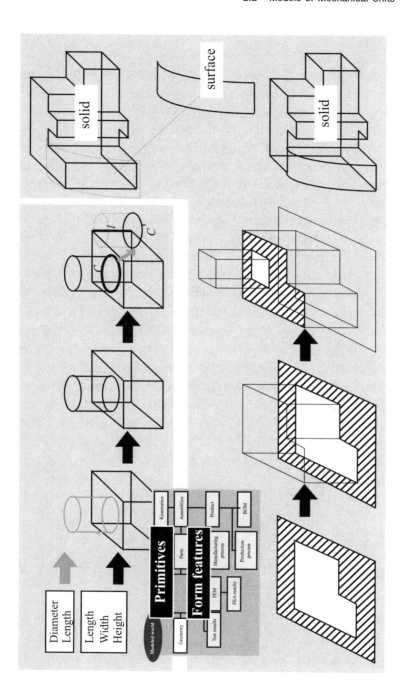

Figure 2-6 Creating complex solid shapes for part models.

systems serves as a construction method and generates a boundary-represented solid.

The second and presently prevailing principle of part shape construction uses modification of an initial shape or base feature by function originated elementary shapes called form features. This construction of the shape is similar to making a sculpture starting from a piece of stone, by removing material, step-by-step to gain details of its final shape. This older method was implemented in advanced CAD/CAM systems during the late 1980s and 1990s. Most of the shape modifications are done by form features defined using a contour sketched in a selected plane on the semi-finished shape of the part.

The lower part of Figure 2-6 explains modeling by form features. A prism solid entity is created as a base shape by tabulation of a closed contour along a height vector. The contour is sketched in the final place on the part. This method is called sketch in place. The base feature is contour based. The rectangular shaped contour for the next feature is defined in the top plane of the base feature. The tabulated feature is generated as an integrated element of the shape by the definition of new topological entities and by mapping new geometric model entities to them. The main difference between the two methods is that element combination uses separate primitives, while shape modification features are created as organic elements in the structure of part boundary. An appropriate sequence of shape modifications by appropriate form features defines the final shape of the part.

The unique world of complex surfaces needs unique modeling methods and tools that are not integrated in solid modeling. Surfaces are constructed separately from the solid part model by separate modeling tools. Ready made and validated surface models are integrated into the boundary of the solid model. Special methods are available to include separate surface models in solid primitives and shape modifications. The most advanced integration allows construction of surface at its final place in the boundary of a solid. The basic concept of the integration of a surface model in a solid is shown on the right of Figure 2-6. In this example, intersection of a surface with a solid modifies the geometry to accommodate the surface in the boundary.

Traditional and versatile computer descriptions of parts are applied for analysis, manufacturing planning, and presentation purposes. Computer modeling in engineering relies upon sophisticated part models and construction methods. However, the part design and manufacturing centered modeling of the 1970s and 1980s has been dramatically extended to integrated descriptions of all possible engineering objects in comprehensive and well-organized product models.

A large step was the integration of assembly and kinematics design with part modeling. Instead of the construction of individual part models, the connection-related design of parts proceeds in an assembly space where the part is constructed and represented in connection with other parts in the actual assembly. The assembly model consists of a structure of connected parts and a set of relationship definitions for part placement in relation to parts that are connected to it. One of the advantages of this method is that connections of parts are defined by form features common for two or more parts.

An advanced approach to assembly modeling is its full integration with part modeling. Assembly modeling is done in an assembly model space where parts are modeled to define shapes and dimensions that are affected by the assembly. Additional details of the part are modeled in the part model space. Part models can be created for an assembly in accordance with their place in a predefined product assembly tree. This is the top-down approach. Existing part models can be used to create a product assembly tree. This is the bottom-up approach. Engineering practice mixes these two approaches because some parts are available at the start of the modeling of a mechanical system while other parts are to be modeled on the basis of their place in the assembly tree. For multiple applications of a part model, it is not duplicated but referred from the assembly model.

The assembly structure is completed by relationship definitions between mating parts such as surfaces in contact, axes in

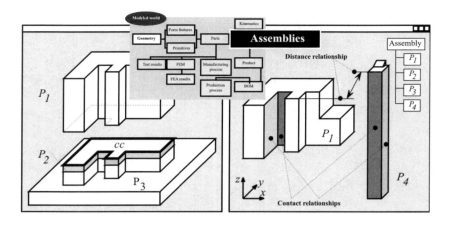

Figure 2-7 Relating parts in the model space.

coincidence, etc. In Figure 2-7, part P_4 is placed on the part P_1 by use of two contact relationships at the shaded flat surfaces and one distance relationship between the appropriate edges. The left part of Figure 2-7 shows an example of shape associativity. The same closed contour is applied as the construction element of solid form features on three different parts. In other words, contour CC is used as one of the parameters of entity creating procedures for parts P_1–P_3. Placing part P_4 relative to part P_1 defines a fixed position of the part in the assembly model.

Parts in a mechanical unit are fixed or moving during its operation. Under the effect of input movements, parts must move in directions specified by engineers and must not have the ability to move in any unwanted direction. This is made possible by unit-wide coordination of degrees of freedom defined individually at pairs of mated parts. Each part has a given number and type of degrees of freedom in relation to a mating part. In Figure 2-8, part P_4 can move linearly in the direction of coordinate z: we say it has one degree of freedom.[1] Allowed movements between

---

[1] A body may have six degrees of freedom at the maximum. Three linear and three rotation movements can be allowed.

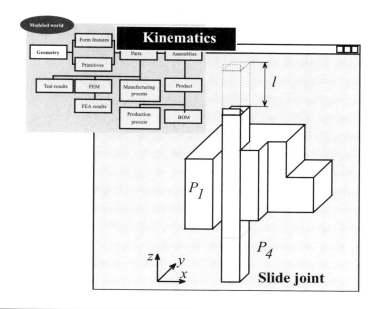

Figure 2-8 Description of kinematics by joint.

appropriate axes, surfaces, etc., of parts are described in the kinematic model. The modeling of kinematics needs information about the connection of parts in the assembly model; thus the assembly model is completed by the kinematics model. In the description of kinematics, parts are considered as rigid bodies and possible movements are described between pairs of parts by joint entities. Figure 2-8 illustrates a slide type joint where prismatic parts P_1 and P_4 slide on each other. The distance of this one degree of freedom linear movement is limited to l.

Parts are being exposed to increasingly severe and complex loads during their recent history of development. To cope with new requirements, materials are constantly being improved and decreasing quantities of highly engineered and expensive materials are being used in parts. These conditions, together with increased emphasis on quality assurance and product liability, have urged the development of sophisticated and effective computer methods for analysis of stress, deformation, heat, vibration, etc., along parts. Finite element analysis has replaced tedious and time-consuming calculations in conventional analyses. The increased power of computers has facilitated the application of finite element methods in the everyday work of engineers. Finite element methods rely on the calculation of analyzed parameters at arbitrary points of a part by use of an appropriate solver. They are an approximation but are much more accurate and reliable than conventional analyses. The application must be harmonized with the problem.

Finite element methods are implemented in Finite Element Modeling (FEM) and Finite Element Analysis (FEA) software tools. FEM builds finite element models while FEA uses these models to solve various analysis problems. FEA sometimes is called finite element solving. FEM is a sequence of modeling activities to prepare geometric models for analysis. It involves modification of the part geometry, finite element mesh generation, placing loads and restraints on the part to communicate real working conditions with the FEA solver, and checking the finite element model. FEA software is dedicated to one or more typical problems such as static or dynamic, that is linear or nonlinear. Calculations use material property information.

Part geometry is modified to build an effective mesh. Some details are eliminated and additional geometric elements are included. The modified geometry is used only for FEM/FEA purposes; manufacturing planning is done by using the original part shape. A choice of finite elements is available for FEM in the library of the modeling system. A finite element mesh can be generated along a line, in a plane, on a shell, or in a solid, according to the shape to be analyzed. A finite element is a structure of nodes and edges. Linear or curved, even fifth degree curved, edges can be applied. At the same time, computer resources and time are saved by approximation of curved part contours by line segments such as edges of finite elements when the analysis task allows this. The density of the mesh is selected to define the simplest appropriate mesh. In practice, serious loading conditions often concentrate on relatively small areas of the part. An initial coarse mesh can be

generated to identify high and low areas for the examined parameter. The coarse mesh is refined by adaptive meshing that
controls the density and other characteristics of the mesh. Loads
and boundary conditions such as restraints are placed on points,
along edges and curves, or on surfaces and sections of the part
geometry. They can be placed also at groups of nodes of finite
elements. Displacements of the part are restricted by the restraints
as effects of the parts bounded with it. Springs, dampers, and
gaps are often involved in FEM models. The finite element
model is checked for consistency and much more; a consistent
model includes a complete net with complete elements as well
as all loads and boundary conditions that have emerged during
the operation of the analyzed part.

The reviewing of analysis results by engineers is assisted by
post-processing. Displacements, magnetic fields, stresses, strains,
temperatures, and other parameters are communicated with engineers by use of color coding, tabulated and filtered numerical
values, or graphs. In Figure 2-9, a ribbed section of a part carries
a mesh consisting of solid linear tetrahedron finite elements. The
expected value range of the FEA results is divided into four sub-
ranges with a color code for each of them. Finite elements are
colored according to the actual values of the analyzed parameter.

The part model is the source of shape information for the
programming of numerically controlled (NC) machine tools. NC
is a flexible machining technology where cutting tools are moved
under computer control to give the designed shape of the part. The
cutting tool generates the demanded shape by moving along its
part model information driven path. In Figure 2-10 volume V_1 is
removed by the moving tool t along tool path Tp_1.[2] A solid geometric model entity representing V_1 carries information on the
material to be removed to create step St on the part P_1.

A machining task is defined as following a contour, creating a
surface, or removing some material (Figure 2-11). Tool movements

---

[2]In other words, the tool trajectory or tool track.

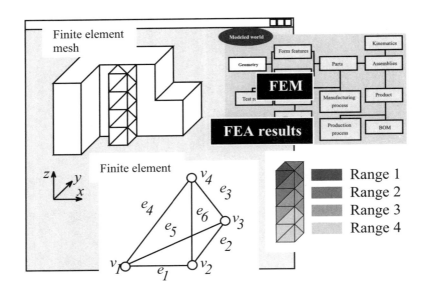

Figure 2-9 Finite element modeling.

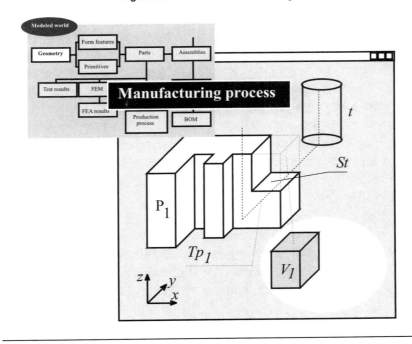

Figure 2-10 Part geometry driven machining.

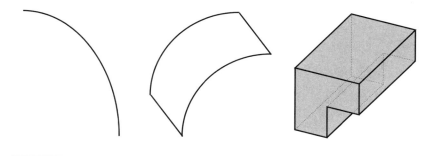

Figure 2-11   Geometry as a machining task.

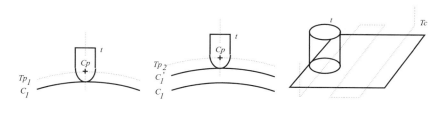

Figure 2-12   Creating tool paths.

are created for these tasks using contour, surface, or volume to be removed entities, respectively. The tool path is the route of the controlled point of a cutting tool. The machined shape can be changed simply by changing the control program according to a change of the part model.

Three situations in the geometric model driven creation of tool paths are illustrated in Figure 2-12. The controlled point C_p of the tool t is moving along the tool path Tp_1. Curve C_1 is a geometry model entity on a part to be created. The situation changes when the curve to be created C'_1 is not the same as the part model entity C_1. C'_1 is calculated from C_1 using a rule, for example, using the relation between the ready made and rough shape of the part. Curves C_1, Tp_1, C'_1, and Tp_2 are related by associativity definitions in the model. Changes of curves in the part model can control changes of tool paths through associativities. If a machining task is defined as a surface to be produced (rightmost part of Figure 2-12) or a volume to be removed, a continuous series of tool paths called a tool cycle is created.

2.3   Integrated Modeling

Engineering modeling uses a large number of proven methods for the description of solid, surface, part, assembly, kinematics, analysis, manufacturing process, and other engineering objects. From the 1960s to the early 1980s, stand-alone modeling software packages served each purpose from the above list. Most human effort was wasted in learning different user interfaces, modeling philosophies, and data formats. Model data of different formats and contents were exchanged between separate databases. Serious difficulties challenged engineers in the exchange of model data between modeling packages with different information description capabilities, different sets of entities, and different parameters to describe the same engineering objects. This situation motivated developments for integrated modeling solutions. Integration efforts were supported by unification and standardization of models, databases, and data communication formats. Systematic, cooperative, and successful development of leading CAD/CAM software products allowed the establishment of globally integrated CAD/CAM systems on the Internet. Figures 2-13 and 2-14 tell the history of this development and summarize its different stages from the 1970s to the present. Because typical demands from different sectors of industry are very different, existing engineering modeling systems apply all of the stages, depending on their purpose, technical level, and environment.

Modeling procedures are collected in software packages. Packages are offered as stand-alone program products or modules of one or more program products. They are toolkits for a well-defined set of purposes. When a CAD/CAM producer composes a toolkit, demands by prospective customers, needs of typical engineering tasks, and requirements of software engineering are considered.

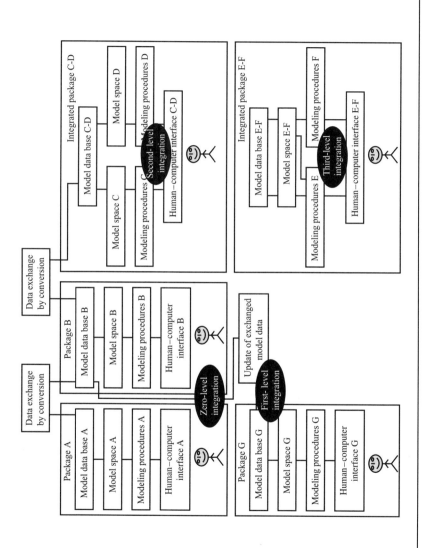

Figure 2-13 Integration of modeling packages.

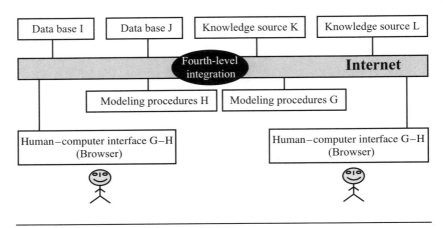

Figure 2-14   Integration on the Internet.

For an easier survey of integration, the authors have defined its five levels, explained in Figures 2-13 and 2-14. On the *zero level of integration*, stand-alone program products (Packages A and B in Figure 2-13) are interfaced. They have individual and different databases, modeling procedures, and user interfaces for the same and different purposes. An engineer who works with two or more packages in the same computer system should have different knowledge and thinking about interactive modeling, the work of modeling procedures, and database content. Model data exchange needs translators for conversions.

An additional problem with data exchange besides data conversion is frequent change of models. The *first level of integration* offers a solution to this problem (Packages G and B in Figure 2-13). Using a special communication, earlier exchanged model data can be upgraded in the received package by the sending package, during the life of the exchanged model.

In the *second level of integration*, a common database and human–computer interface integrate modeling procedures in a modular software environment where standard communication is established amongst modeling procedures, the database, and the human–computer interface (integrated package C-D in Figure 2-13).

Sometimes two different modeling activities can be done in the same model space, then the result distributed to several model spaces. This is why *third-level integration* is defined as an upgrade of second-level integration by integration of model spaces (integrated package E-F in Figure 2-13).

The most important application of third-level integration is in modeling parts in an assembly context (Figure 2-15). *Parts B* and *C* are created in their model space; they carry contours to be copied on to *Part A*. The solution is creating *Part A* in the assembly space after *Parts B* and *C* are placed in it. Other details of *Part A* are

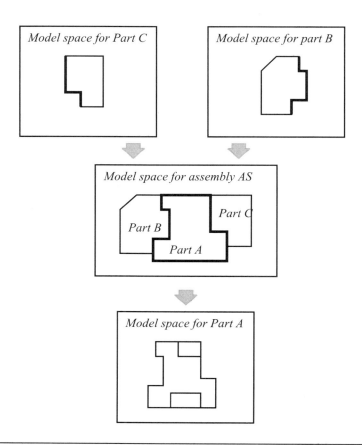

Figure 2-15 Modeling of parts for assembly.

completed in its model space. This modeling requires the integration of part and assembly modeling and the related model spaces.

Fourth-level integration is based on the Internet. Sets of modeling procedures, databases, and knowledge sources are placed at various computer systems on the Internet. Special purpose browsers manage access of authorized engineers working at sites authorized to access the above listed resources.

Integration ensures the collaboration of engineers and easy access of product data in cooperating modeling systems. Model data are managed in heterogeneous computer and multi-modeling environments. Multi-modeling is a means for modeling with several different systems in engineering of the same product or family of products. The recent trend is to apply product information in its original form in order to avoid any loss of information at conversions and to retain integrity of the model. On the other hand, the product model STEP[3] of the International Standards Organization (ISO) was developed for model data exchange without loss of any information.

Often several engineers work on the same product design and create surface, part, assembly, analysis, production tool, and tool path models simultaneously. A downstream application can start when the minimal necessary information is made available by upstream activities (Figure 2-16). This *concurrent engineering* provides reduced time and early recognition and notice of an inappropriate or erroneous principle, concept, design, method, object, attribute, or process before detailed design of the product. Concurrent engineering relies upon relationship definitions in the model towards downstream applications.

2.4 Entities

One of the most frequent questions from engineers, at their first attempt of computer modeling, is: what is the real meaning of the

---

[3]Standard for Exchange of Product Model Data, ISO 10303.

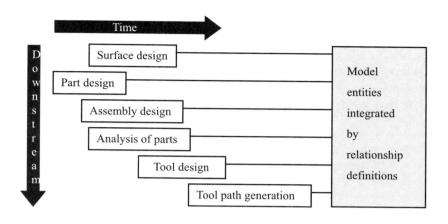

Figure 2-16   Simultaneous engineering.

word *entity*? It is not easy to define this term. A possible definition is that an entity is an elementary, a complex, or a structural unit in a model that is generated by a dedicated model-creation procedure and can be distinguished from any other entities by its type and parameters. Entity describes an engineering object or some feature of it. There is an actual list of entities in the model data pool at any time during modeling sessions. Entities are stored in the database and exchanged with other modeling systems as model components. Higher-level entities use lower-level entities at their definition as parameters of model-creating procedures. Each modeling system has the capability to handle a specified set of entity types. In the case of incompatibility between two modeling systems, an entity exchanged from a sending modeling package may be illegal, non-understandable, or non-processable for a receiving modeling system. Sometimes one of the parameters of an understandable and processable entity causes incompatibility. There is a real danger of quality problems when the model is processed without validation.

There are several general expectations of the handling of entities in modeling packages. A modeling package is expected to include all necessary procedures to create, combine, modify, and

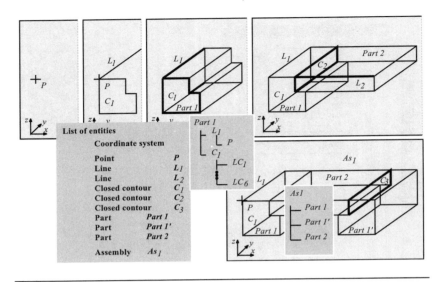

Figure 2-17 Entities in an assembly.

delete the specified entities. Each allowed activity during a model-creation session must be reversible at any time during and after the modeling session. If any deletion of an entity results in broken associativities or destroyed integrity, the modeling system must send notice of repair information for the engineer.

Figure 2-17 summarizes the general characteristics and construction of shape model entities. Let us suppose that a model creation starts with an empty model space. The first entity is point P as defined by its coordinates. Both line L_1 and closed contour C_1 entities are defined using P as the starting point. We say that L_1 and C_1 are higher-level entities than P. Contour C_1 consists of six straight-line entities. Any component element of C_1 can be modified or deleted and new elements can be inserted. Entities L_1 and C_1 are input parameters for the definition of a solid prism entity. This solid is the only component entity of the solid model representation of *Part 1*. Contour C_2 is taken from *Part 1* for the definition of *Part 2*. Line L_2 is defined as an extension of an edge of *Part 1*. Similarly to *Part 1*, the shape of *Part 2* is represented by a single solid prism. The prism is defined using L_2 and C_2 as its parameters. Assembly As_1 consists of *Part 1*, *Part 2*, and *Part 1'*. *Part 1'* is a mirror copy of *Part 1* and is related to *Part 2* by contour C_3 opposite to C_2 on *Part 2*. Existing entities at this point of model construction are listed in Figure 2-17. Lower-level entities for the creation of higher-level entities *Part 1* and As_1 are also listed.

Shape entities are visualized for their interactive creation, modification, and application in a vector graphic handled region of the screen called a view port. As in Figure 2-17, lines and contours generally carry enough information for model construction. If not, simple or realistic filling, shading, and light source definitions visualize surface entities.

An entity depends on entities in the lower or the same level of entities. Manual modification of all dependent entities on the same and lower levels is time consuming, hard to survey, and a source of errors. Additional information for relationships between pairs of entities in the model description makes automatic propagation of modifications possible. This model information relates parameters of the same entity or different entities as associativity. Associativities integrate stand-alone entities into an associative model. Figure 2-18 illustrates the effect of modifications without and with associativity definitions in the model. In Figure 2-18a, closed contour C changes to C'. Tabulated solid *Part A* is generated using C' as a construction curve. When the associativity definition is included in the model between C and *Part A*, any change of C is propagated to *Part A*. In Figure 2-18b, the same situation is explained for entities on the same level. Change of *Part A* can be propagated to *Part B* by associativity definition between appropriate dimensions between *Part A* and *Part B*.

According to their function in the description of engineering objects, relationships can be represented between model entities as simple relationships, parameters, references, shape modifications, common shape modifications of several parts, variants, combinations, and structures (Figure 2-19). A *simple relationship* relates parameters of entities. In the example of Figure 2-19a, dimensions

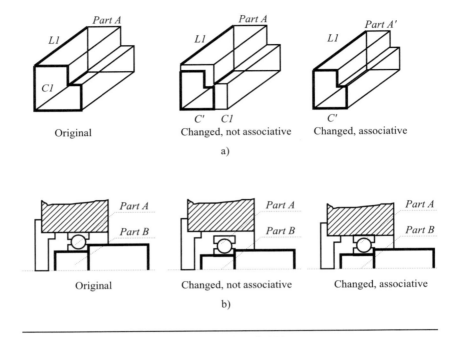

Figure 2-18   Associativities.

a and *e* on *Part 1* and *Part 2* are related. Another example of a simple relationship is a contact assembly relationship between surfaces of these parts. Entities e_1 and e_2 are *parameters* of entity e_3 in Figure 2-19b. Entity e_{r1} serves as a common *reference* (datum) line for entities e_1 and e_2 in Figure 2-19c. Entity e_{r2} acts as a reference plane for entities e_3 and e_4. Entity e_2 is created by *modification of the shape* of e_1 by form feature *FF1* in Figure 2-19d. Form feature *FF2* extends to *Part 1*, *Part 2*, and *Part 3* in the *Assembly* in Figure 2-19e. A variant is defined on the basis of given differences from other entities. Entity e_5 is a variant of entity e_2, the difference is the mirrored position of a shape modification step (Figure 2-19f). *Element combination* is done by Boolean operations such as union and subtraction of solid primitive entities. Entity e_2 is created as union of entities e_6 and e_7 (Figure 2-19g). A *structure* defines relations amongst a set of entities as parts and subassemblies in an assembly model in Figure 2-19h.

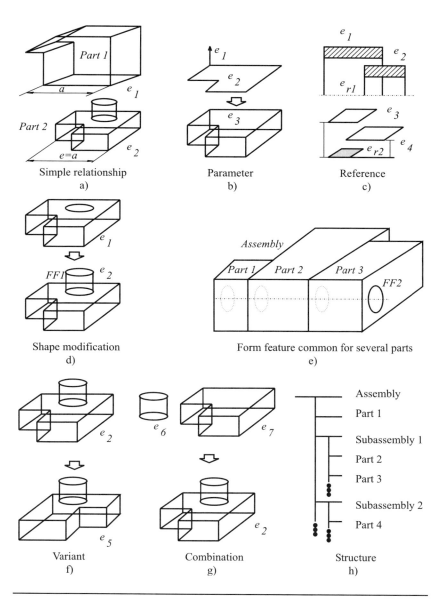

Simple relationship
a)

Parameter
b)

Reference
c)

Shape modification
d)

Form feature common for several parts
e)

Variant
f)

Combination
g)

Structure
h)

Figure 2-19   Relationships between entities.

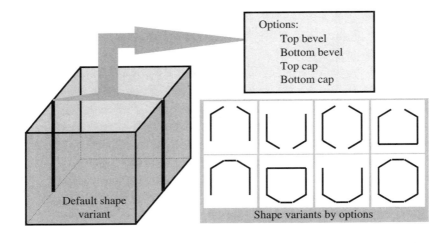

Figure 2-20 Shape options.

To avoid a huge number of shape types, modifications of a default type by use of option definitions is applied. Figure 2-20 shows a surface with open or closed ends. The surface is created using its section contour as a construction curve. The section is configured according to the shape variant to be created by use of additional *optional elements* such as a bevel and cap on both of its ends. The section is modified according to the selected optional elements. In the example of Figure 2-20, combinations of four options produced eight additional shape variants.

2.5 Open Architecture Modeling Systems

CAD/CAM systems are well-organized, structured, integrated, and comprehensive tool sets for all necessary design functions. However, it is impossible to develop a CAD/CAM system that fits exactly to all possible tasks at all industrial companies even in a relatively narrow area of application. CAD/CAM systems are *customized* to extend their standard capabilities to meet special customer needs for their applications. Open architecture features

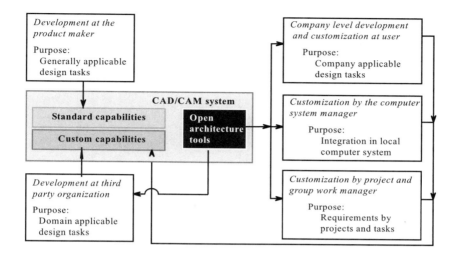

Figure 2-21   Development of CAD/CAM systems.

facilitate the easy customization and development of CAD/CAM systems in their application environments. User interface customization, macro language capability, direct data access, and direct command and function access are the main methods for programming modeling systems at their application.

Figure 2-21 outlines the scenario of development in open architecture modeling systems. The original developer of the program product develops *generally applicable design procedures* such as *standard capabilities* and *open architecture procedures* as tools for the development of *custom capabilities*. The main benefit is that development can be done where the domain expertise is available. Authorized third party software development organizations add application domain related elements to CAD/CAM products. Customization at a company installation is done on *three levels*. The first is the level of *company related development* for standards, etc. At the second level, the manager of the computer system customizes the CAD/CAM system for *integration into the local software environment*. Finally, *group work and project managers* customize the system on the third

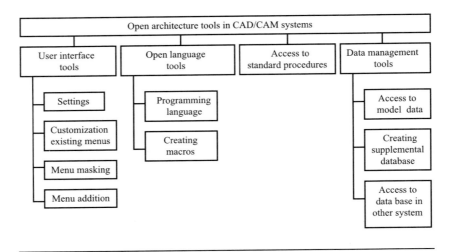

Figure 2-22 Open architecture tools in CAD/CAM systems.

level to meet local design tasks and personal requirements of engineers.

Open architecture software tools are available in modeling systems for user *interface customization, programming, standard procedures access, and database handling purposes.* Figure 2-22 outlines the basic functionality. Programming languages are developed to be easy to learn and apply by engineers. Company-specific data sets and databases can be developed and remote databases can be accessed by programming using open architecture tools. Third party and user developers can continue the development of the program product without the need for any special product software related knowledge.

Open architecture developments result in external programs. All fundamental programming functions, such as definitions of local and global variables, logic expressions, loops, compare operators, and mathematical expressions are available. Menus for these programs are integrated into the existing user interface. The visibility of commands on the user interface can be controlled by masking according to demands from the functionality of the installation. Macros can be captured during modeling sessions and custom macros can be written for repeated applications.

There are several basic developments, such as post processors, STEP data exchange models using the EXPRESS language, and programs for Internet applications. The importance of accessing remote model databases and modeling systems has gained special attention during recent years. User definitions of mathematical expressions and rules are cited in this text many times. Launched applications can be run without leaving the CAD/CAM system environment, and the user interface of outside systems can be emulated.

Customization is considered a part of the installation such that modeling systems without some customization features cannot cope with the demands of present industrial practice. A customization toolkit is considered to be an essential structural unit of industrial modeling systems.

Computer Representations of Shapes

Up to this point, elementary methodology and basic considerations have been discussed. Chapters 3–6 outline model representations available for the solution of engineering problems. Entities such as elements of engineering object representations in models of mechanical systems in present industrial practice are explained in Chapter 3. Then integrated modeling is considered by discussion of associative elementary representations and their structures in Chapters 4–6. Product modeling, as a recent idea for highly integrated, group work of engineers, as an advanced form of concurrent engineering and human–computer interaction, and as a means for the description of design intent is also given. The following are steps in the process of model entity definition.

Definition of engineering objects to be modeled.
Identification of model entities.
Definition of entity attributes.
Definition of relationships between entities.
Placing entities in structures.

These steps determine the information content of model entities. Model representation must serve the objectives and desired applications of the information stored in models.

3.1 Geometric Modeling in a Nutshell

This section is not intended to be a detailed explanation and discussion of mathematics as applied in geometric modeling. Instead, it is a means of connecting mathematics with engineering modeling. Following this, the characteristics of curves and surfaces for the description of engineering shapes to solve problems from present modeling practice are discussed and explained. Details of geometric modeling such as the phenomena, creation, and manipulation of mathematical equations are given in books dedicated to these aspects.

3.1.1 Boundary Representation

The boundary of a shape is composed of *surfaces*, as well as *curves* as intersections of surfaces. Complex shapes consist of a large number of curves and surfaces. Both the outside and inside boundaries can be defined on mechanical parts. The following are fundamental goals of the boundary type of shape representations.

All curved and planar surfaces as well as curves and lines must be described by using the same class of mathematical functions (Figure 3-1a).

Surfaces must have closed contours or curves in their inner and outer boundaries (Figure 3-1b).

At the connection of surfaces, a specified continuity must be kept. In Figure 3-1, surfaces SC_1 and SC_2 are connected with a specified value of tangency along their intersection curve (Figure 3-1c).

The shape must be closed at connections of surfaces. This requires that all points of any common border curve must lie in both of the connected surfaces (Figure 3-1d).

The boundary representation must allow all necessary geometric operations during construction, modification, and
application of the model such as intersection by curves
and surfaces, and modification of continuity conditions.
Additional information is necessary for mapping surface–
curve–surface connections. Boundary representation
applies the traditional mathematical method of topology
for this purpose.

The boundary model is based on the recognition that a closed
shape is bounded by a set of topological faces. A face is related to a
region of the part surface. Each region is described by a closed and
orientable surface. A face is bounded by edges and vertices so that

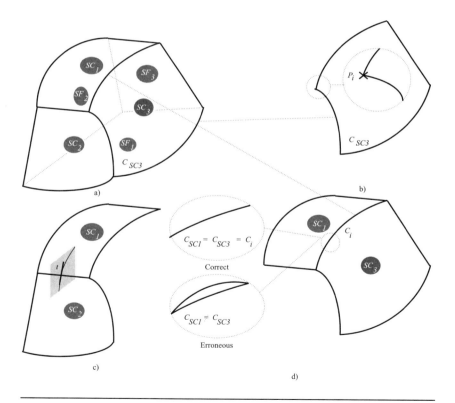

Figure 3-1   Boundary representation of a shape.

they are appropriate for the connection of intersecting surfaces. An edge is related to an intersection curve or line. As a result, the geometric model for boundary representation is composed of two interconnected groups of entities, namely topological and geometrical entities (Figure 3-2). The main difference between the two groups of entities is that topological entities are connected in a structure and do not carry any shape information, while geometrical entities are stand-alone entities and carry only shape information. Elementary topology and geometry are given briefly in Figure 3-2.

Solid shapes having all planar surfaces, such as a cube or a box, can be handled as polyhedral. If the shape of the covering surfaces is left out of consideration, the surfaces can be replaced by shapeless faces. The shape is mapped to the face in the form of a surface geometric entity. Application of topology in shape modeling utilizes this recognition. Information about the intersection curve common to the two surfaces is attached to an *edge* topological entity. Edges are connected by common *vertices* at their connections. A *face* is bounded by a closed chain of edges and vertices. The result is a *consistent topology* that accepts mapping of all geometrical information in the form of descriptions of curves and surfaces.

Topology is not easy to understand for engineers who worked previously with conventional engineering drawings and pictures. The visualized model in the view port of the screen can be misleading for the engineer who is not experienced in the construction of geometric models. As an example, a cube may be recognized by an engineer on the left of Figure 3-3. It seems to be a visualization of a body and the flat surfaces in its boundary seem to fit along borderlines. However, this figure visualizes a model consisting of six independently defined flat surface entities in a model space. The model does not contain any information about connection of surfaces. The evidence of this is a successful attempt of translation transformation of single surfaces S_1, S_2, and S_3 along vectors v_1, v_2, and v_3, respectively. An engineer seeing this visualization on the screen may consider the model to be a complete and consistent

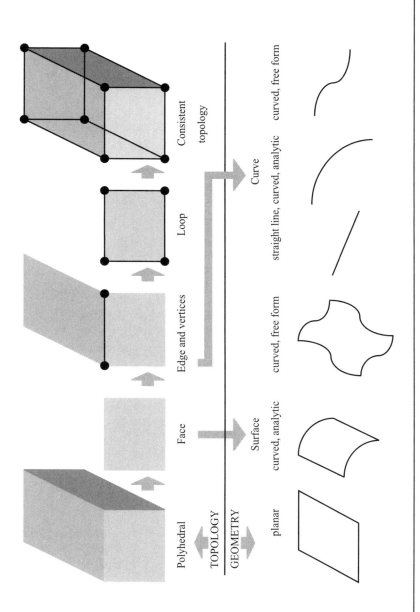

Figure 3-2 Topology and geometry.

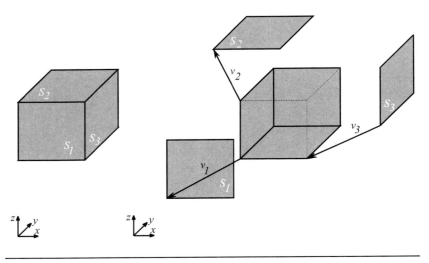

Figure 3-3 Surfaces in the boundary of a shape.

one for a solid. Similar misunderstandings may lead to serious problems.

Another misunderstanding involves application of *constructive solid geometry* (CSG). In early solid modeling, CSG was a leading model representation; now it has been replaced by boundary representation. CSG is still important as one of the model construction methods for solids, but CSG modeling procedures generate boundary representation.

Boundary description is the fundamental and prevailing concept of contemporary shape modeling. Advanced shape modeling systems utilize the advantages of geometrical-topological structures for engineering and other purposes. It should be noted that certain old modeling systems still in application might not able to do topology. To facilitate the processing of boundary models in these systems, data exchange procedures extract curves and surfaces from the shape model. Although lost topology information makes the model hard to look at, pure geometric information is enough for some purposes, for example for calculation of tool path information on part geometry. Engineering practice accepts *intended and controlled loss of information.*

3.1.2 Geometry

The shape of an engineering object may be composed of *predefined, controlled,* and *free form* elementary shapes. On the other hand, geometric elements are *linear* and *curved.* Predefined shapes can be described by simple mathematics so that they are called analytical shapes. Linear analytical shapes are lines and flat surfaces. Curved analytical shapes are conics, cylindrical surfaces, cones, tori, and spheres. Circles and ellipses are the most common conics in engineering. Other conics are parabolas and hyperbolas. The form of predefined shapes is fixed. Any other shapes can be altered as controlled or free form. Controlled surfaces are created by surface generation rules such as tabulation, rotation, or sweeping. Free form shapes are free form curves and surfaces. They may have arbitrary shape; however, their initial shape must be defined by curves or points for the procedures that generate them.

Geometry in a boundary is a set of elementary or complex surfaces with elementary or complex curves at their intersections. An elementary surface is created using curves and other parameters as a stand-alone entity. Complex surfaces are generally blends of two or more elementary surfaces. Blending is a method to remove edges at connections of surfaces, making a single or multiple surfaces. A complex surface carries information about its original components for later modifications.

Examples of primitive and complex surfaces can be seen in Figure 3-4. Curves C_1 and C_2 are input parameters for creation of surface S_1. Elementary surfaces S_1 and S_2 are connected by blending into surface SC. Curves C_1–C_6 are arranged in a curve network for the creation of component surfaces CS_1–CS_4.

Figure 3-5 illustrates how surfaces build up the geometry in a boundary. The geometry of boundary B_1 consists of four flat and two curved surfaces, as well as intersection lines and curves. Flat surfaces are defined by trimming. This operation is a restriction of a surface by a closed chain of lines called a trimming curve. S_1 is a surface primitive and is created by using two curves applying one of the basic rules of surface creation. Surface S_4 is created by

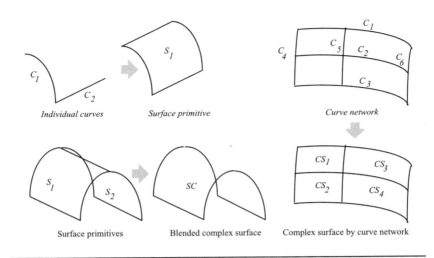

Figure 3-4 Primitive and complex surfaces.

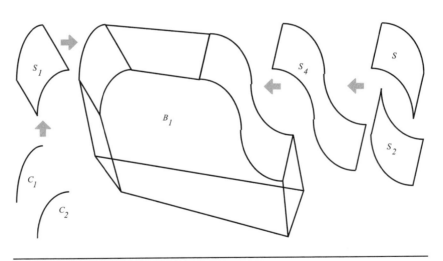

Figure 3-5 The geometry concept of a boundary model.

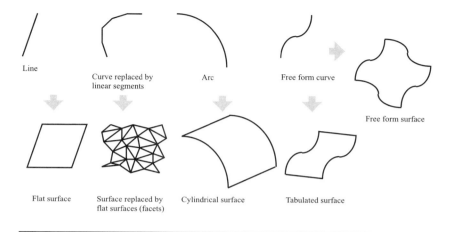

Figure 3-6 Geometrical entities.

blending surfaces S_2 and S_3. S_2 and S_3 are created using the same rule as at S_1. The continuity specification may need modification of component surfaces at their boundary.

Figure 3-6 summarizes typical groups of lines, curves, and surfaces. Lines and flat surfaces can be described by very simple mathematical functions, and the handling of linear geometry is still very beneficial despite the increases in computer performance. Curves and surfaces can be linearized by replacing them with polygons and sets of small flat surfaces called facets, respectively. This is beneficial when the decreased demand for computational performance of linear geometry ensures feasibility of the modeling task or enhances the efficiency of finite element analysis, visual reality, control of production equipment, and rapid prototyping. Although limited computer performance as a reason for handling tasks as linear is over, the economical application of computer resources takes the advantages of linearization. Control of the tool path in machining, visualization of surfaces by computer graphics, and rapid prototyping are inherently linear.

Simple curves can be defined by an analytic shape such as an arc that serves as a construction curve for a cylindrical surface in Figure 3-6. An increasing percentage of parts need more complicated shapes in the form of controlled or free form geometry. Sometimes it is enough to define a free form curve, then apply it for the creation of a surface by a simple control rule. In Figure 3-6, a free form curve is moved along a straight line defining a tabulated surface. A free form curve can serve as one of the border curves of a free form surface. Free form curves and surfaces are created as a synthesized effect of several functions. They are often cited as synthetic surfaces.

Free form curves and surfaces are described by polynomials in engineering practice. Among other advantages is that their derivatives can be calculated easily. In early curve and surface modeling, analytic shapes were represented exactly by simple functions or approximated by polynomials. Special polynomials were developed for the exact description of analytic shapes during the 1980s. Now, *unified geometry* applies polynomials for the description of all shapes in parts. The old wish of the application of a single function for all shapes has been fulfilled. Modeling uses the same type of mathematical function and data structure for analytic, controlled, and free form curves, surfaces, and trimmed surfaces.

Construction of boundary representation using elementary and complex surfaces would be complicated and time consuming; instead, engineers define solid form features. Application related shape concepts are captured in boundary represented form features as illustrated in Figure 3-7. The first step is creation of base feature *BF*. The upper plane surface *P1* of the base feature *BF* serves as a base to define two form features FF_1 and FF_2. The base contour C_1 is created in the plane P_1. The *sketch in place* construction method guarantees that all points of C_1 lie in the plane P_1. Contour C_1 is used at creation of form feature FF_1. It is integrated into the boundary of the shape. S_1 represents the case where the surface is created outside of the solid model by separate surface modeling tools. It is projected onto the plane P_1. A volume feature is created between S_1 and the projection of its boundary. In this way, it is integrated into the boundary of the shape.

At a connection point of two curves (Figure 3-8) or along a connection (common) line or curve of two surfaces in the boundary

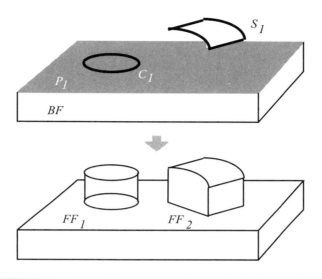

Figure 3-7   Definition of form features using geometrical entities.

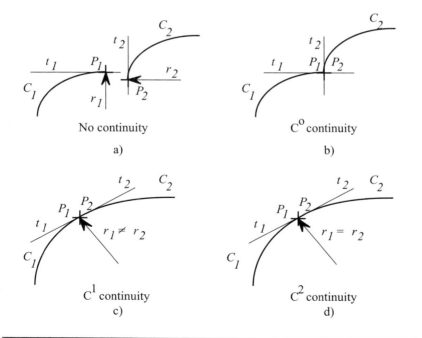

Figure 3-8   Continuity.

of a shape, a specified continuity must be kept at the creation and modification of the connected entities. Different denominations are applied for levels or, in other words, degrees of continuity. The usual denomination of the actual level of continuity is the letter G and the number of the level (e.g., G1) or the letter C with the level number in as a superscript (e.g., C^1). In Figure 3-8a, curves C_1 and C_2 have no common point, their end points P_1 and P_2 are not congruent, the complex curve is discontinuous. Zero order continuity of curves C_1 and C_2 (C^0) yields their common point (Figure 3-8b) but other criteria of continuity are not fulfilled. In the case of first order continuity (C^1) the slopes of the curves (first derivatives or tangents) are the same at their connection point (Figure 3-8c). Continuity C^1 is often cited as a first derivative or tangent continuity. The tangent vector is represented by a continuous function. When second derivatives or curvatures at the connection point of curves C_1 and C_2 are the same, the continuity is of second order (C^2, Figure 3-8d).

Specification of continuity may be an inherent characteristic of a model, determined by an existing shape model environment, or predefined by the engineer (Figure 3-9). Model-creation and modification procedures handle specification as constraints and controls of the continuity accordingly. When shapes that are more complex are composed using single entities and original shapes that do not provide the specified continuity, one or more components should be modified. Curve and surface entities, especially the imported ones, may contain discontinuities. Analysis recognizes and repairs continuity problems before application of the model.

3.1.3 Topology

As explained above, topology describes *neighborhood* information for geometric entities. While geometry is tangible, topology is abstract. Note that vertex, edge, and face topological entities are symbolized by bullets, lines, and shaded rectangles, respectively.

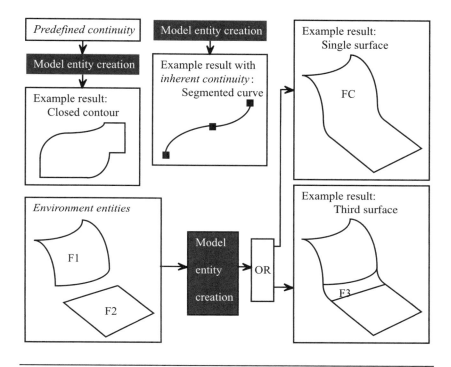

Figure 3-9   Control of continuity.

These elements in this case act as symbols and do not represent any shape.

The basic construction of topology is very simple. Where a *point geometric* entity is defined at the connection of line or curve geometric entities, a *vertex topological entity* is placed in the model (Figure 3-10). The point geometric entity is mapped to the topological entity. Where the *line or curve geometric entity* connects surface geometric model entities, an *edge topological entity* is defined. Consistent topology requires vertices on both ends of an edge. A *face* is enclosed by a closed chain of edges and vertices; this is the *loop* topological entity. It is mapped to a closed boundary contour of a surface. However, a closed boundary contour does not describe geometry; geometric information is stored at its component lines and curves.

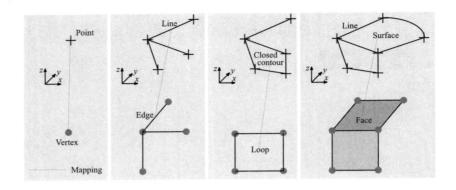

Figure 3-10   Joint definition of topology and geometry.

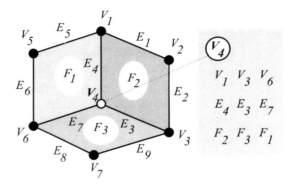

Figure 3-11   Neighborhood in topological structure.

The neighborhood definition describes relational information between topological entities (Figure 3-11). Mapping of geometrical entities to topological entities ensures that neighborhood information describes relational information between geometrical entities too. A *vertex* at the end of an edge is common for several edges. Edges running into it, vertices on the other end of those edges, and faces at the edges constitute its neighborhood. Topological structure is described in the form of a matrix tree or graph using neighborhood information. Figure 3-11 shows the *neighborhood of topological vertex V_4* as represented by a matrix.

Different shapes can be described by the same topology if they have the same number of surfaces and intersection curves. As an example, Figure 3-12 shows three different shapes with the same topology. Modification of a surface or curve often requires modification of curves and surfaces that are mapped to entities in the neighborhood of the topological entity to which the modified geometric entity is mapped. In Figure 3-13, four different

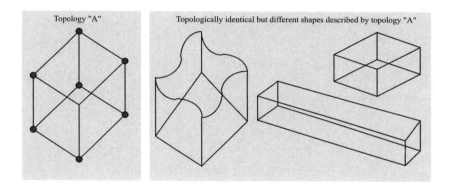

Figure 3-12   Topology and geometry.

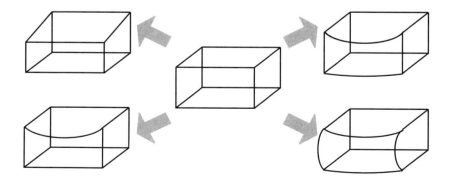

Figure 3-13   Several topology assisted modifications of a shape.

changes of the same geometry are achieved without any change of topology. They are shortening a line, replacing a straight line with a curve, and changing a flat surface to a curved surface with two and four curved boundary lines. To maintain closed geometry, geometric entities must be modified in the neighborhood of the actual topological entity.

The boundary representation of a shape is often composed by using entities resulting from different procedures and sources. Attention should be given to possibly incomplete or erroneous topology or geometry. Measures for checking completeness should be taken after the main stages of model creation. Completeness of the topology is called topological consistency. Consistent topology involves all entities needed to fulfill the objectives of modeling. The topologies in Figures 3-14a and 3-14b are considered consistent ones for solid and wireframe models, respectively. Figure 3-14c illustrates a model in which a topological face

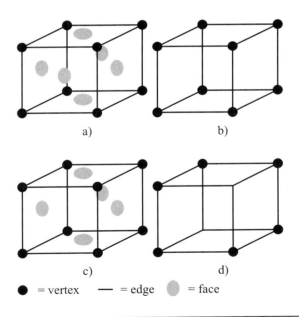

Figure 3-14 Consistent and inconsistent topologies.

is omitted, resulting in inconsistent topology. However, this model is considered as consistent if controlled omitting of topological entities is allowed. Note that an incomplete model is always a special case. Engineering modeling generally requires a complete topology. The topology in Figure 3-14d is inconsistent by a fatal error. Omitting a vertex is not allowed because this action breaks the chain of the fundamental topological structure and makes the connection of geometrical entities unrecognizable. Remember that wireframe models do not accept faces because their geometry does not involve surfaces.

The model of a body requires full consistent topology in the boundary model. This consistency must be created and maintained during all topology related developments and modifications of the model. Fortunately, simple topological checks are available. They are based on four simple rules for consistent topology (Figure 3-15a):

Vertices must be defined on both ends of an edge.
More than two edges are to be run into a vertex.
A face must be bounded by a closed chain of edges.
An edge must be connected to two faces.

Figure 3-15b shows examples of inconsistent topology.

Consistent topology of a boundary always can be projected to a sphere from inside, with no intersection of edges. These are called manifold topologies. Non-manifold topologies break basic topological rules. Two typical non-manifold topologies are shown in Figure 3-15c. On the left, an edge is connected to three faces. However, the third face is sometimes allowed for construction purposes but not as an element of the boundary topology. The related geometry is considered as a mixed application of three and two dimensions in a single model. On the right of Figure 3-15c, two manifold topologies are connected by a common vertex. They constitute non-manifold topology.

Leonhard Euler proved for polyhedral and polyhedral-like objects having non-self-intersecting faces and closed orientable

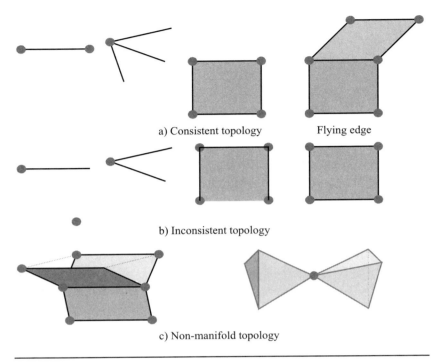

a) Consistent topology   Flying edge

b) Inconsistent topology

c) Non-manifold topology

Figure 3-15 Simple topological rules for complete boundaries.

surfaces that

$$V - E + F = Constant,$$

where V, E, and F are the total number of vertices, edges, and faces in the boundary, respectively. This is *Euler's law*.

$$\chi = V - E + F$$

is the Euler characteristic of the boundary. For a boundary without inner loops on its faces, separate bodies, or through holes, the value of Euler's characteristic is

$$\chi = V - E + F = 2.$$

Consistent topology satisfies Euler's law. In Figure 3-16, the consistency of three topological structures is checked by Euler's law. The central structure consists of three faces, three edges and two

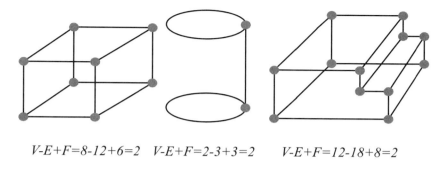

$$V\text{-}E\text{+}F\text{=}8\text{-}12\text{+}6\text{=}2 \quad V\text{-}E\text{+}F\text{=}2\text{-}3\text{+}3\text{=}2 \quad V\text{-}E\text{+}F\text{=}12\text{-}18\text{+}8\text{=}2$$

Figure 3-16 Examples of Euler's law.

vertices. It describes shapes similar to the cylinder. The loop with a single vertex is symbolized by a circle.

In the case of shapes having inner loops on faces, separate bodies, and through holes, the *Euler–Poincaré law* can be applied. It is an extended version of Euler's law:

$$V - E + F - R = 2(S - H),$$

where V, E, F, R, S, and H are the total number of vertices, edges, faces, inner loops on faces, separate bodies, and through holes, respectively.

Evaluation of the geometric consistency of a model is not geometry but model development and application process oriented. It is based on four criteria:

Understandable entities.
Closed boundary.
Shape as it was intended.
Continuities in accordance with specification.

All curve and surface entities in a consistent geometry are known for all modeling systems involved in its development and application. The model must be developed for its applications in accordance with the model representation capabilities of the receiving procedures. Some conversions of the models are allowed, when the result saves all the information carrying the original design intent for its applications.

A closed boundary includes surfaces with no breaks or unintended holes as well as gap-free connections of surfaces along boundary lines and curves. Surface representation without topology does not include neighborhood information for connecting curves and surfaces. Consistency is not guaranteed by simple topology. Although boundary curves at the connection of surfaces S_1 and S_2 in Figure 3-17a are created using the same curve, this does not guarantee gap-free connection. Surface creation procedures often result in different boundary curves on connecting surfaces even though they are mapped to the topology. The solution is a common boundary curve of the two surfaces. The introduction of an additional topological entity, the *common edge*, among others, offers a solution to this problem. A single curve is mapped to the common edge. All points of this curve must lie in both of the connecting surfaces.

A topological problem solved by the common edge is illustrated in Figure 3-17b. Because this edge is adjacent to two faces, it is a component of two loops. These loops have different orientations so that the edge must have twin orientations. The topological entity of the common edge is split into two sides; each side features the orientation of the actual loop. The common edge is also called the split edge. The common edge, or

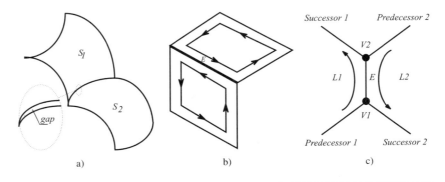

a)   b)   c)

Figure 3-17 Common edge.

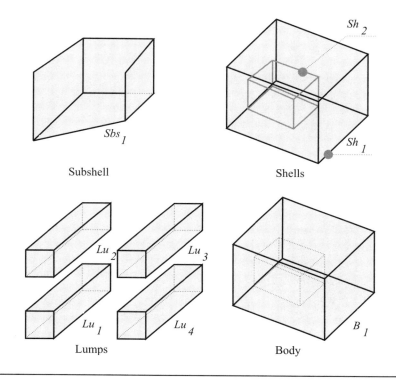

Figure 3-18   High level topological entities.

by its abbreviated name coedge, is built into the topology by the winged edge structure (Figure 3-17c). Predecessor edges run into the common edge, while successor edges run out of it.

Handling bodies in modeling systems requires *higher-level* topological entity definitions. They are groups of vertices, edges, loops, and faces (Figure 3-18). The complete "closed" topology of a boundary for a body is represented by the entity *shell*. Euler's law is defined for checking the topological consistency of a shell. Details of a boundary are defined as representations of form features in advanced part models. The topological entity *subshell* (sbs_1) serves this purpose. The shell entity describes the topology of both outside or inside boundaries (Sh_1, Sh_2). The body is a topological entity representing a solid. The orientation of the surfaces answers the question of where the material is. A group

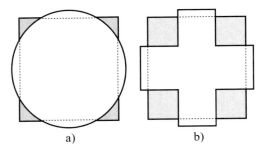

Figure 3-19   Creation of lumps by intersection of bodies.

of separate bodies is defined as a unit where the component bodies are to be kept and handled together. In this case, the group is a *body topological entity* (B_1) while its components are *lump* topological entities $(Lu_1–Lu_4)$.

The primary application of lumps is the description of segments. Figure 3-19 shows two typical examples of the creation of lumps by the intersection of two solids. In Figure 3-19a, a cylinder is cut by a prism. When the diameter of the cylinder is larger than the dimension of the base rectangle of the prism, the result is four separated solids. In Figure 3-19b, lumps are created by the intersection of two prisms; the body topological entity includes four lumps.

Euler operators are tools for the creation of consistent topology for boundary representations. Based on the theory of *plane models* they place *Euler primitives* in the topological structure. A plane model is a mathematical abstraction of the boundary model and a well-proven tool for the evaluation of topological properties. The creation of a plane model is illustrated by the example of a cube in Figure 3-20. The faces in the boundary are separated by separating the two sides of the common edges. The separated loops are represented in a plane by their vertex and edge topological entity components. The plane model is a *planar, directed graph* because it can represent the orientation of edges.

A simplified representation is the condensed plane model. It is a plane model where the edges are not separated. Modification of a

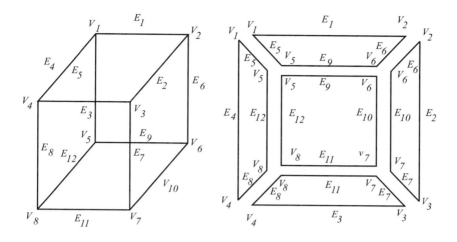

Figure 3-20 Plane model.

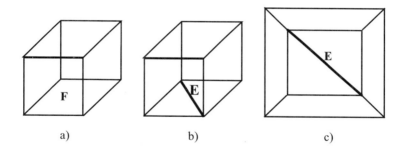

Figure 3-21 Condensed plane model and its modification.

condensed plane model is illustrated by inserting edge E in the topological structure in Figure 3-21. Edge E divides the bottom face *F* into two faces.

The Euler operator is *local* or *global* according to the extent of its effect on the topological structure. A local Euler operator acts on edges or faces, whereas a global Euler operator combines two shells into a single one, divides a shell into two components, or creates a hole.

Figure 3-22 explains how basic local Euler operators build topology in boundary models. An operator is abbreviated by the

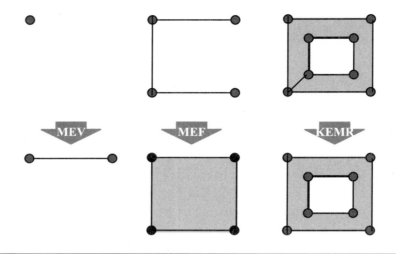

Figure 3-22 Local Euler operators.

initial letters of the actions it does. In a valid topology, an edge has vertices on both of its ends. This is why an MEV operator makes (M) an edge (E) then a vertex (V). An MEF operator creates an edge (E) to close a loop then a face (F) is originated from the loop. Opening a face by an inner loop (ring) requires making a connecting edge and a vertex by an MEV operator. Following this, four MEV operators create an inner loop. At this point, the connecting edge is deleted and a ring is created by the KEMR: kill (K) edge (E) make (M) ring (R) topological operator.

Figure 3-23 shows applications of MEV and KEMR operators. A shape is modified by the extension of its boundary with the topology and geometry of a form feature representation (Figure 3-23a). Surface S_1 is selected in base feature BF as reference surface for creating form feature FF_1. The topology for S_1 was created on BF as shown in Figure 3-23b. The face for S_1 is opened by the application of the MEV operator for a connecting edge, four MEV operators to create inner vertices and edges, then a KEMR operator to make the inside ring (Figure 3-23c).

Creating the topology for a complete shell is illustrated in Figure 3-24a. It starts with a one-vertex one-polygon model

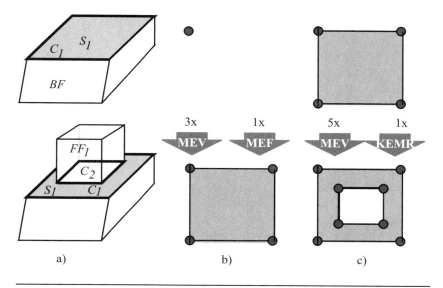

Figure 3-23   Combined application of local Euler operators.

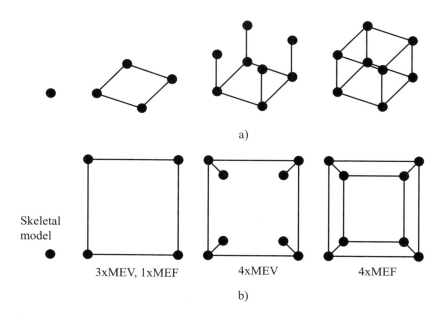

Figure 3-24   Creating topological structure of a cube using local Euler operators.

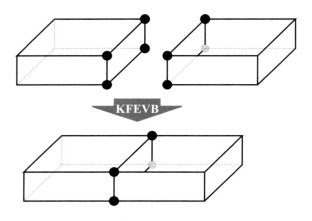

Figure 3-25 Connecting topologies by the KFEVB operator.

called a skeletal model. Starting from the skeletal model, three MEV, one MEF, followed by four MEV, and finally four MEF Euler operators are applied. The development of the plane model for the topology can be followed in Figure 3-24b.

One group of the higher-level topological operators combines or separates bodies. The KFEVB operator kills a face and associated edges and vertices as well as the body carrying them. The result is a fusion of two bodies (Figure 3-25). Other combinations of difference and intersection are similarly supported by Euler operators. Higher-level operators create bodies for primitives such as cubes, cylinders, and spheres.

3.2 Representation of Geometry

This section continues the initial discussion of the computer description of geometry. The objective is to deliver knowledge to understand the principles, methods, and practice of industrially applied shape modeling in CAD/CAM systems. The reader can find detailed discussion of computer-represented geometry in

specialized books.[1] The present text focuses on model characteristics for the application of geometry in problem solving for
industrial engineering practice.

3.2.1   Basic Requirements

Industrial products include mechanical parts and mechanical parts
are also included in the equipment, devices, and tools for their
production. The shape of a part is designed, analysed, and manufactured according to a specification considering particular
requirements from the following areas:

> customer demand
> forecast
> type of product
> standards
> available production processes
> financial conditions
> application environments
> business strategies
> legislation.

Part design relies upon advanced shape modeling. The application of high performance shape modeling methods can be beneficial only if all specifications are fulfilled. The introduction of
application orientations broke down the barrier between the high
level theory and application of shape modeling during the 1990s. A
computer model of a shape must support the description of the
characteristics of a part as specified for downstream engineering
activities. This determines the required modeling capabilities of
part modeling systems for the creation, analysis, modification,
and application of curve and surface models. The main criterion

---

[1]McMahon, C., and Browne, J. "CADCAM – From Principles to Practice,"
Addison-Wesley, Reading, MA, 1993; Zeid, I. "CAD/CAM Theory and
Practice," McGraw-Hill, New York, 1991.

is to create and maintain consistent geometry. Normally a consistent description of curves and surfaces must support the following modeling activities:

All geometric calculations.
Creation of all derived curves and surfaces.
Any modification.
Trimming by appropriate curves.
Intersection with any curve and surface.

Aesthetic, mechanical, electrical, aerodynamic, hydrodynamic, and manufacturing related requirements of curved surfaces are specified according to the application of the part to be modeled. Specifications include shape, dimensional, surface quality, and tolerance related characteristics. As the carrier of the most frequently communicated information between different modeling systems, the geometric model is expected to support all foreseeable data exchanges with and processing in the receiving modeling system.

3.2.2 Curve Representations

Advancements in curve modeling for engineering purposes opened a new era in engineering design during the 1970s and 1980s. In this subsection, a brief historical introduction is followed by a discussion of the characteristics, properties, and features of curves. Curves are focus objects in engineering design:

they are input objects for the creation and trimming of surfaces;
they connect surfaces in boundary representation;
they carry information about continuity at interconnections of surfaces;
they define actions on shape;
they are applied at the definition of finite elements; and
they define paths and trajectories for the control of production equipment.

3.2.2.1 Brief History of Curve Representations   The computer description of curves for industrial engineering purposes is connected to the name of Paul Bezier. He worked for the car firm Renault in France and developed curve description based on Bernstein polynomials during the late 1960s and early 1970s. Bezier introduced a control polygon for the control of the shape of curves at their creation by approximation. Vertices of the control polygon acted as control points; the shape of the curve is controlled by the movement of control vertices. The results of Bezier's work were first applied in the historical *Unisurf* surface modeling system at the Renault car factory in France in 1972. Because the primary goal in the design of the outside surfaces of the body of a car is rather a more harmonic than accurate shape, approximation (Figure 3-26b) was more appropriate than interpolation (Figure 3-26a). An interpolation

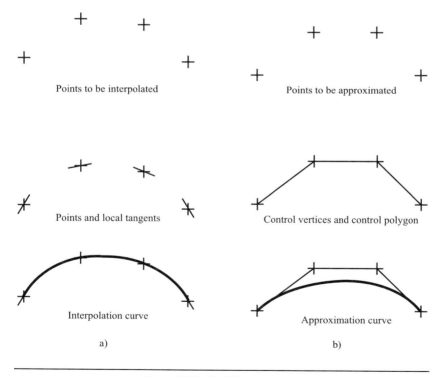

Points to be interpolated

Points to be approximated

Points and local tangents

Control vertices and control polygon

Interpolation curve

Approximation curve

a)

b)

Figure 3-26   Definition of curves.

curve is defined by the points which the curve is forced to go through. Recently, the creation of high order interpolation curves has produced highly engineered and styled shapes. As a result, both approximation and interpolation curves are applied in the modeling of styled shapes. The shape of an approximation curve is modified by moving *control vertices*. The interpolation curve is modified by moving its *drag points* or by a change of direction of its tangent at selected points. The drag point can be an interpolation or any other selected point of the curve. The tangent may be specified directly, controlled by associativity, or given by manual graphic interaction. First order continuity at connection points of the curve and other entities can also be defined by tangents.

Approximation is only a method of harmonic shape control by the position of control vertices of a control polygon. Decreasing the distances between the curve and the control points is not a goal. Approximation replaced the troublesome curve definition by point-tangent pairs at that time.

A similar curve description was achieved by Philip De Casteljeau at the French firm Citroen. Nevertheless, the method of approximation of the control polygon was linked to the name of Bezier in the literature even if the description of the curves uses functions other than Bernstein polynomials.

A Bernstein polynomial-driven Bezier curve is a one-piece curve. It goes through the first and last points of its control polygon and is tangential to the first and the last polygon segments (Figure 3-27a). The Bernstein polynomial attached to the control points acts on the entire curve; consequently, the displacement of a control point changes the entire curve. This is the global type of curve control. The degree of the curve is the number of control points minus one. The class of the curve equals the number of control points. On closing the control polygon by a line between the first and last control vertices, the inside of the resultant closed polygon is a convex hull (hatched region in Figure 3-27b). All points of the curve are within the convex hull.

When a large number of control vertices was approximated, the degree of the curve was high because of the application of

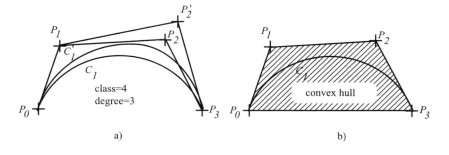

Figure 3-27 Basic characteristics of a Bezier curve.

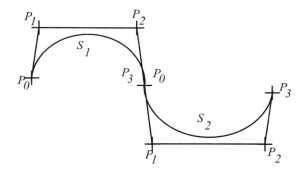

Figure 3-28 A chain of two Bezier curves.

Bernstein polynomials. High degree Bezier curves were difficult to handle: they were often unstable in engineering applications. To solve this problem, high degree Bezier curves were replaced by a chain of lower degree Bezier curves. In the example of Figure 3-28, a curve of six degrees is replaced by two curves of three degrees. Continuity has to be ensured at the connecting point of the curve segments. Advanced Bezier curve representations are still widely applied in industrial practice.

As a solution to the problems of Bezier curves, the B-spline curve gained a leading position during the 1990s. The B-spline curve is inherently segmented, in other words piecewise. The degree of the curve is the same as the degree of the polynomial and it does not depend on the number of control points. Application of B-spline geometry solved the problem of unstable high degree polynomials. All analytic and synthetic shapes can be described by rational B-spline functions. Analytic shapes are not approximated: they are exact. Now, rational B-spline functions are applied primarily as representations of geometry. Unified geometry, where all shapes are described by the same function, relies on rational B-splines. Note that there are also rational Bezier curves.

The spline function is a mathematical model of the spline, a traditional tool of shipbuilders to create curved edges on the ship. The physical form of the spline was a flexible metal strip. This strip was spanned between fixed points on a base material then a smooth curve was drawn along it.

3.2.2.2 Characteristics of Curves

Industrial practice of shape modeling applies only a small number of representations for curves. *Polynomials*[2] are the preferred class of mathematical functions for the description of curves and surfaces. *Basis functions* are often called *blending functions* because they affect the shape of the entire curve (*global control*) or only several of its segments (*local control*). They are connected to control vertices or interpolation points.

Modeling activities of a curve require the identification of curve points in a way that is independent of the position of the curve in the model space. For this purpose, a *parameter* is associated with the curve. The parameter should have a continuously increasing value along the curve, from its starting point to the end. The *parametric representation* of the curve is its *parametric equation*. It relates model space coordinates to the parameter value at a

---

[2]The reader will find a more detailed discussion of blending functions and other mathematics of Bezier and B-spline curves in: Gerald Farin; *Curves and Surfaces for Computer-Aided Geometric Design*, Academic Press, San Diego, 1997; Anand, V. B. "Computer Graphics and Geometric Modeling for Engineers," Chapter 10, John Wiley & Sons, Inc., New York, 1993.

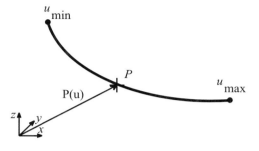

Figure 3-29 Parametric representation of curves.

given point of the curve. The parametric representation is applied almost exclusively for curve and surface modeling in present day engineering practice. The position of a point on the curve is given by the position vector (Figure 3-29)

$$\mathbf{P}(u) = [x(u)\, y(u)\, z(u)],$$

where $u_{\min} \le u \le u_{\max}$ and $\mathbf{P}$ is the position vector of point P in the model space. This is the general form of the parametric equation of the curve. Model space coordinates at point P can be calculated as a function of parameter u:

$$x = x(u), \quad y = y(u) \quad \text{and} \quad z = z(u).$$

The functions, principles, methods, and details for mathematical calculations are hidden from the engineer during the construction of the geometry by general-purpose industrial modeling systems. However, basic knowledge of the following characteristics of curve geometric model entities is essential for engineers:

Definition of curves by approximation or interpolation (Figure 3-26).
Exact analytic or free form shape.
Convex hull (Figure 3-27).
Parametric description (Figure 3-29).
Blending functions.
Degree and class.
Local characteristics (Figure 3-30).

One-piece or piecewise (composite) curve.
Global or local control of shape.
Continuity at ends and segment borders.
Start and end point conditions.
Method of parameterization (Figure 3-31).
Weight at control or drag points.
Orientation.

Modification of a curve is global when any movement of any control vertex changes all points of the curve. In local modification, movement of one control or drag point affects the shape of a segment of the curve. Segments connected to the modified segment may also be modified to maintain the specified level of continuity at segment borders.

Calculations of tangent, normal, and curvature values at different points of curves and surfaces are basic operations in geometric modeling and need mathematical functions with good derivability as polynomials. Evaluation of polynomials requires only multiplication and addition of real values. The general form of a polynomial function of degree n is:

$$p(x) = a_n x^n + a_{n-1} x^{n-1} + \cdots + a_1 x + a_0,$$

where n is a non-negative integer and $a_0, a_1, \ldots, a_n$ are real numbers. The simplified general form of polynomials is:

$$p(x) = \sum_{i=0}^{n} a_i x^i.$$

The position of the curve for any value of the parameter is the sum of the values of the blending or basis functions. In other words, the resultant curve is a combination of influences of blending functions. The effects of blending functions are "blended." When a basis function has influence on the entire range of the parameter, the control of the curve is global. The value of basis functions is non-zero at any value of the parameter. Global modification of Bezier curves is the result of the application of Bernstein blending functions. When the influence of a basis function is for only a limited range of the parameter, the control of the curve is local. Segmented B-spline curves with spline blending functions show local modification. A B-spline curve can be considered as a generalization of a Bezier curve.

The *global characteristics* of the curve such as its *control*, *degree*, and *class* are defined for the entire curve. Calculations during creation, modification, and application of a curve use the following *local characteristics* (Figure 3-30):

value of parameter (u);
tangent (t);
normal (n);
binormal (b); and
curvature (c).

The local characteristics t, n, and b define the accompanying trieder. The tangent and normal planes are defined by t, n and b, n, respectively.

The possible level of continuity at the connection of segments depends on the degree of the blending functions. Cubic B-spline curve segments have gained the widest application in practice because they ensure second order continuity at their connection points. The shape of linear or quadratic analytic curves cannot be modified to maintain the curvature continuity at segment borders after local modification of the curve. Recent well-engineered and styled B-spline curve representations require ten and even higher degree curves.

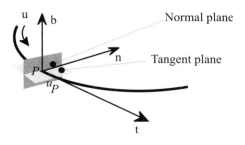

Figure 3-30   Local characteristics of a curve.

Parameterization of piecewise or segmented B-spline curves needs special attention. A distinct parameter range is defined for each segment. The connection point of two adjacent segments is called a *knot*. The parameter value valid at a segment connection is attached to the actual knot and is placed in the *knot vector*. Consequently, a knot vector carries information about parameter ranges assigned for segments of a B-spline curve. The knot vector for a curve consisting of *m* segments of degree *n* is:

$$U = \{u_0, \ldots, u_i, \ldots, u_m\}.$$

Figure 3-31 shows a curve consisting of four segments. In this example, the parameter range of the curve is divided equally for the segments. An interval u_i, u_{i+1} is called the *i*th *span* of the curve. The entire parameter range of the curve can also be divided unequally amongst segments. Decreasing the parameter values along a knot vector is not allowed:

$$u_i \leq u_{i+1}.$$

When parameter intervals along the knot vector are equal, the B-spline curve is *uniform*:

$$\{123\}.$$

A B-spline curve represented by a knot vector in which parameter subrange values are repeated, is *periodic*. It can be

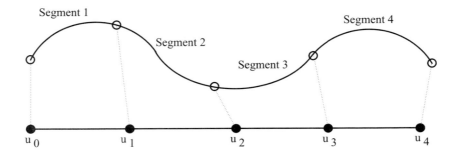

Figure 3-31   Segmented (piecewise) curve.

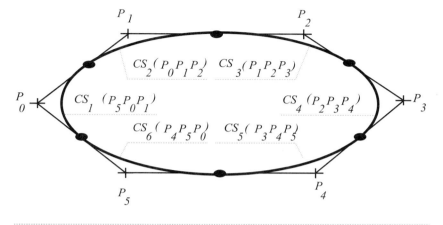

Figure 3-32   Closed B-spline curve.

seen that a *uniform* B-spline curve is also periodic. A uniform B-spline curve can be closed. For instance, the curve in Figure 3-32 consists of six segments (CS_1-CS_6). Each segment is under the control of three control points; two of these are common to neighboring segments. Modification of a segment by moving a control point also modifies the previous and the following segments in order to maintain continuity at segment borders.

The B-spline is *non-periodic* when parameter ranges are equally distributed for inner segments, but intervals are repeated at the beginning and the end of the knot vector. A span may have the same parameter range as the previous span so that knot values are repeated. These are *non-periodic* curves. The number of repetitions is called the *multiplicity*. The maximum allowed number of repetitions is the class of the curve. A zero parameter range is allowed for a span. Examples of knot vectors for non-periodic B-spline curves of different degrees with the maximum allowed multiplicity are as follows:

$$\text{linear}: \{001233\},$$
$$\text{quadratic}: \{0001222\},$$
$$\text{cubic}: \{00001111\}.$$

It is easy to recognize that the cubic curve in the above example is a Bezier curve. A Bezier curve can be considered as a special case of a non-periodic B-spline.

When parameter intervals along a knot vector are different, the B-spline curve is *non-uniform* as in the following example:

$$\{0,0 \;\; 0,2 \;\; 0,5 \;\; 0,6 \;\; 0,9 \;\; 1,0\}.$$

As explained above, the *rational B-spline curve* is suitable as a representation of all analytic, controlled, and free form shapes. It provides a unified mathematical description of the geometry for boundary representations and surface models. The word *rational* comes from the construction of the function as an algebraic ratio of two polynomials. Both Bezier and B-spline curves have a rational form. The prevailing representation in present geometric modeling is the *non-uniform rational B-spline* (*NURBS*). The NURBS has been accepted as an international and national standard. It has several key advantages over non-rational curves for everyday modeling practice:

Invariance for projective transformations.
Exact description of analytic shapes such as lines, circles, and other conics.
Excellent local modification.
Unlimited number of control points.
Degree of curve as high as ten or more.

The rational curve is based on the principle of homogeneous coordinates. Detailed and application oriented discussion of mathematical issues for rational curves is available in specialized books.[3] The main essence of homogeneous coordinates is given below.

A point in *Euclidean space* is:

$$P(x, y, z).$$

---

[3]For a detailed discussion of the mathematics and application related problems, see Piegl, L., and Tiller, W. "The NURBS Book," Springer, Berlin, 1997.

It can also be represented in *four-dimensional homogeneous space* as:

$$Q^w = (wx, wy, wz, w),$$

where $w_i \geq 0$ is the homogeneous coordinate. This is also called the *weight* and is attached to control or drag points of the curve for the control of the shape. By changing the weight values at control points, the same rational function can describe different analytic shapes such as line, ellipse, parabolic, and hyperbolic. Actual values of the weight for control points are included in the *weight vector*. Each of the control vertices has an initial weight at the creation of the curve. The weight can be changed as a modification of the curve. When all control vertices or drag points of a curve have the same weight, the curve is non-rational.

In conclusion, methods for the control of rational B-spline curves are as follows:

Position of control or drag points.
Specification of continuity conditions.
Values in knot vector.
Values in weight vector.

3.2.3   Surface Representations

3.2.3.1 Well-engineered Surfaces   In the past, simple outside shapes were typical on products, because the paper based medium of engineering drawing, the available shape manufacturing capabilities, and the high cost consequences limited the feasible shapes. At the same time, some products received their shapes from manual shape-adding procedures. Intricate curved shapes were created in the form of *hand-made molds* and *templates* for machining. Fantastic shapes were manufactured during the nineteenth and early twentieth centuries. When electronic and later computer control of machining was introduced, engineers were faced with a problem: there was no available information about the shape as input for control. At the same time customer demand forced the introduction of advanced shapes for consumer and other industrial products. Styling moved slowly into engineering design. The market success of modeled shapes resulted in a breakthrough in the advancement of engineering related styling during the 1990s. The solution was shape model based design in computer systems and capturing hand-made shapes as computer models, by reverse engineering. Now, the human imagination for shape is limited only by the capabilities of mathematical description and manufacturing processes. This is possible at a reduced cost in comparison with conventional design and manufacturing.

The following are features of modeling for well-engineered surfaces:

Full control of the shape is possible during interactive modeling sessions.
Variation of curvature can be controlled along the surface.
High order surfaces can be handled.
Existing physical shapes can be captured and converted into consistent models.
Continuity can be maintained according to specifications.
Model driven manufacturing of shapes can be realized without any loss of information.
Shape can be defined taking into consideration all decided restrictions by manufacturing and economical considerations.

The criteria of well-engineered surfaces are impossible to define because they vary with advancements in shapes and shape modeling processes as well as with shapes preferred by both customers and styling engineers. Flexible shape design and manufacturing offers a fast change of shapes on products required by a new style of engineering.

3.2.3.2 Characteristics of Surface Representations
As with curves, surfaces are described by parametric equations. Two parameters, normally denoted by u and v, identify a point P on the surface (Figure 3-33). The position of the point P is given by the position vector $\mathbf{P}$

$$\mathbf{P}(u, v) = [x(u, v)\, y(u, v)\, z(u, v)],$$

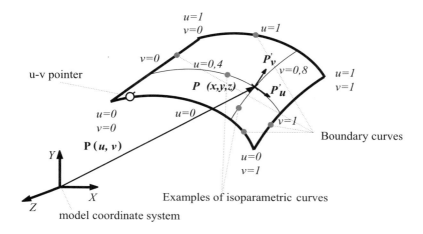

Figure 3-33   Parametric representation of surface.

where $u_{\min} \leq u \leq u_{\max}$ and $v_{\min} \leq v \leq v_{\max}$. This is the general form of the parametric equation of the surface.

The model space coordinates of point P are given by the parametric equation of the curve as a function of the parameters u and v:

$$x = x(u, v), \quad y = y(u, v), \quad z = z(u, v).$$

Figure 3-33 also shows tangent vectors P_u', P_v' of the point P in the direction of the parameters u and v. The value of one of the parameters is constant along isoparametric curves.

The parametric surface must have points for all possible value combinations within the parameter ranges. This condition is fulfilled when any point in the *parameter space* is mapped to a point in the *model space* (Figure 3-34). The concept of parameter space is based on the recognition that parameters can be considered as local coordinates. The parameter "space" is a flat surface enclosed by a rectangle with dimensions equal to parameter ranges.

An elementary surface is created as a known shape; as a shape generated according to a law using arbitrary curves or fit to existing points or curves, depending on the shape and the available input information. A surface as described by mathematical functions has no inherent boundary curves. It can be restricted by

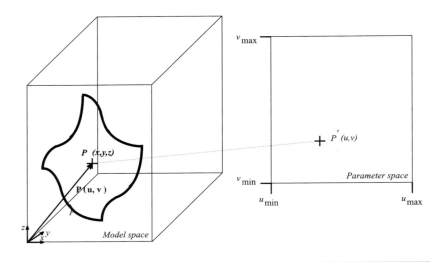

Figure 3-34   Model and parameter spaces.

a closed curve or closed chain of curves called a *trimming curve* during and after its creation (Figure 3-35a). *Trimming* as one of the basic surface manipulation operations creates boundary curves around the useful region of the surface. The original extent of the surface can be restored when it is necessary for model development without any change of the trimmed surface. Some modeling systems in the 1980s and early 1990s could not handle trimmed surfaces; surface models had to be exchanged with application systems together with trimming curves. A surface also can be restricted by intersection with other surfaces.

Elementary surfaces are interconnected by their boundary curves. Complex surfaces are created by adding elementary or complex surfaces to elementary or less complex surfaces. Boundary curves connecting the elements of the complex surface are applied in their original form for creating new surfaces (Figure 3-35b and c), modified according to continuity specification for the connection (Figure 3-35d), or created by intersection of surfaces (Figure 3-35e).

Surfaces, connected along a line or curve, are often merged into a single, smooth surface. The boundary line or curve can

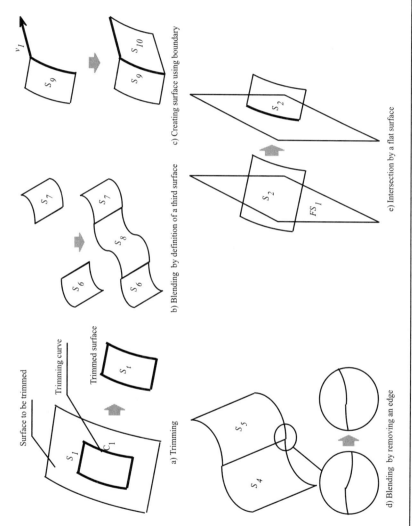

Figure 3-35 Creating complex surfaces.

be removed by modification of one or both surfaces. When two surfaces to be merged do not have a common boundary, a third surface is created with a specified order of continuity at the connections of the new surface with the original surfaces. In Figure 3-35c the *tabulation* rule is applied using a trim edge in the boundary of surface S_9 and vector v_1 as input information. Curves as inputs for the creation of surfaces also can be free curves in the model space or arbitrary curves on surfaces.

Curves on existing surfaces serve as inputs for a variety of construction activities. Any point of a *curve on a surface* lies in the surface and coincides with a point of the surface. Curves on surfaces and other curves defined in relation to surfaces are as follows (Figure 3-36).

When a surface consists of patches, *patch boundary curves* are defined (C_{pb}).

The surface is enclosed by *boundary* or *edge curves* (C_e).

An isoparametric curve is created by fixing one of the parameters (u or v) of the surface (C_{is}).

Intersection curves are defined by cutting the surface with another surface (C_i).

A curve outside of the surface (C_p) can be projected onto the surface from a projection center (P_p) in given direction to generate a *projected curve* (C_p').

Curve C_c is defined by points that go through P_1, P_2, and P_3. The points of the curve as generated by interpolation between pairs of these points are not necessarily points of the surface. When a *curve on a surface* is to be created, points of C_c that are out of the surface must be pulled onto it.

Curve C connects curves C_1 and C_2 with continuity of tangent or curvature at the connection points. Pulling the resultant *connection curve* onto the surface may be necessary.

The *shortest curve between two points* (C_s).

The distance between two curves. It can be defined on the surface (C_{ds}) or in the model space (C_{dm}).

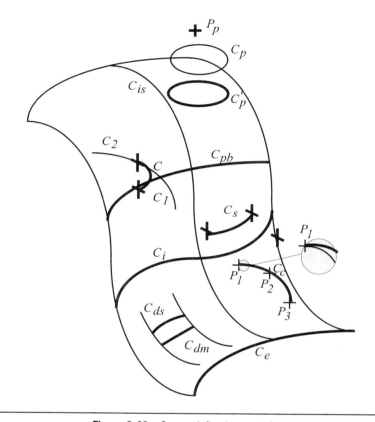

Figure 3-36   Curves defined on a surface.

3.2.3.3 Methods for Control of Shape   The main purpose of shape modeling is to generate different shapes on a part for different purposes, following intentional and intuitive shape related ideas of engineers. Several basic methods are available to control the shape of parts. They are illustrated in Figure 3-37 and summarized below.

A surface may be created by one of the *shape control rules*. As an example, curves $G_1(u)$ and $G_2(u)$ in Figure 3-37a are used to create a surface by definition of straight lines between points of the same u parameter value on the curves.

Points can be *approximated* or *interpolated* to gain a surface. A set of points can be applied to create surfaces directly as

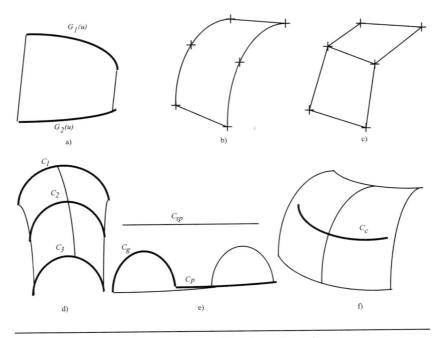

Figure 3-37 Control of the shape of a surface.

interpolation points (Figure 3-37b) or indirectly as vertices in a control polygon network (Figure 3-37c).

Any *curve* applied in the creation and modification of *surfaces* inherently controls the shape. Surfaces can be defined by a single set or two sets of curves to lie on the surface. Creating surfaces by *shape control rules* integrates one or more curves in the surface. For example, the lofted surface in Figure 3-37d contains curves C_1, C_2, and C_3. These curves were used as inputs at the creation of the surface.

Spines and rails control the *shape* of elementary surfaces. At the creation of a swept surface, a generator curve (C_g) is moved along a path curve (C_p). The normal of the plane of the generator curve is defined at each point along the path curve by the tangent of a *control curve* called a *spine* (C_{sp}) (Figure 3-37e). The shape of surfaces can be controlled by curves outside of them. A special curve called a *sculpt curve* controls the shape

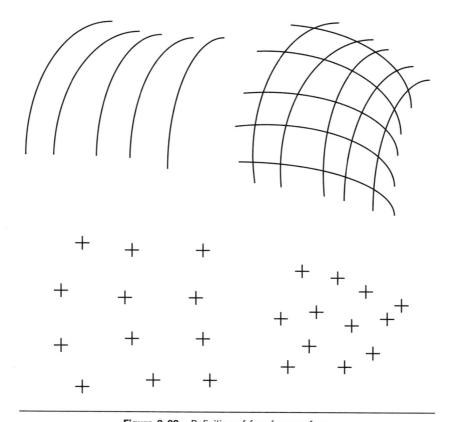

Figure 3-38   Definition of free form surface.

of complex surface C_c in Figure 3-37f. The surface is associative with curves as applied at its definition. Any modification of these curves causes modification of the surface. This is one of the methods to adjust surfaces to their changed or developed model environment.

Free form surfaces, as their name suggests, can have any shape. Variations of their shape are limited by the capabilities of model representation and manufacturing processes. Input information (Figure 3-38) defines their initial shape:

a set of curves;
two intersecting sets of curves;

an array of points;
a cloud of points.

Free form surfaces are also called *sculptured surfaces* because the result is sometimes like a sculpture. These surfaces sometimes are created in a manner similar to making a sculpture. Curve set inputs suggest the third name for free form surfaces: *mesh surfaces.*

Points are *calculated* or *measured.* Curves are calculated or they approximate or interpolate sets of points. Surfaces defined by aerodynamic and hydrodynamic design are created using points and curves defined using calculations or experiments. Shapes of existing physical objects are digitized to gain points. There are several tasks where reverse engineering is the right solution. Styling engineers often prefer to use physical material such as wood or clay at the definition of a shape. They do not like to or cannot put their artistic talent and manual skill into shapes in virtual environments. Reverse engineering can be applied to capture beautiful shapes carried, for example, by old castings and forgings, even wooden sculptured pieces.

Most of the surfaces in part boundaries are created by one of the *shape control rules.* The shape is generated according to the selected rule. Control rules are often used to give an initial shape, and then some regions are modified as free form surface or by associative relations to other parts.

A *flat surface* can be created by use of any known definitions such as two straight lines, three points, or one line and one point. A flat surface is bounded by a closed 2D line, curve, or contour.

A *tabulated surface* is created by projecting a generating line, curve, or compound line ($G(u)$) along a line or a vector (v) (Figure 3-39a). The *surface of revolution* is obtained by rotating a plane curve ($G(u)$) around an axis (Figure 3-39b). Several analytic surfaces, such as a cylinder, torus, or cone, can be defined by revolution. A *ruled surface* is generated by linear interpolation between points of the same parameter values on two curves ($G_1(u)$ and $G_2(u)$) in Figure 3-39c). Directions of parameters for the

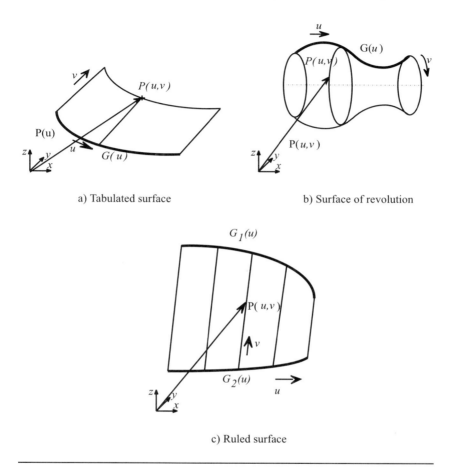

a) Tabulated surface   b) Surface of revolution

c) Ruled surface

Figure 3-39   Simple rules for the creation of surfaces.

surfaces and the position vectors of their points are also shown in Figure 3-39.

The shape control method of *lofting* creates a surface that goes through predefined section curves. In Figure 3-40, section curves C_{s1} and C_{s2} are specified as inputs for lofting. The task is to interpolate them. A large number of shapes can be defined by their interpolation between the two section curves. When a

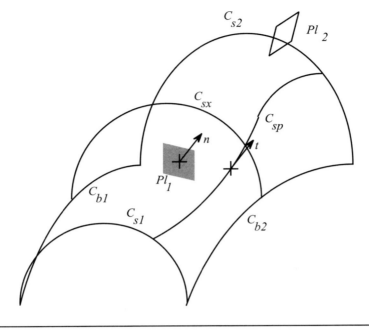

Figure 3-40 Control of shape by lofting.

definition that is more detailed is required, additional initial section curves can be defined. The tangent constraint (t) between the section curves and the spine (C_{sp}) can be applied for additional control of the shape. The effect of the spine on the shape of the surface can be shown at a section curve C_{sx}, created between the two input section curves. The direction of the normal n of the plane Pl_1 of the section curve C_{sx} is the same as that of the tangent t of the spine at its point where the section curve is generated. The effect of the model environment on the lofted surface also can be taken into account. In the example of Figure 3-40, the normal of the plane Pl_2 must be in the same direction as the tangent of the section curve C_{s2} and the surface is restricted by boundary curves C_{b1} and C_{b2}. Parameter u is attached to the direction of interpolation while parameter v is defined along the section curves.

The most versatile control of shape is offered by *sweeping* (Figure 3-41). The surface is generated by a *generation curve G*. It is moved along a *path curve C_p* under the control of a *spine C_{sp}*. The spine controls the direction of the normal of the plane of the generator curve along the path curve similarly as explained in Figure 3-40 for lofting. The generation curve in a given position along the path curve is called a *profile curve (C_{pr})*. Parameter u of the surface is defined along the path curve while parameter v is bounded to the profile curve. Consequently, profile curves are isoparametric curves. Profile curves C_{pr1}–C_{pr3} also can be created as entities associative with the swept surface where the construction of the geometry requires them. Pivot point P_p, or in other

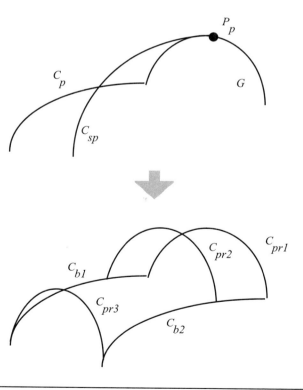

Figure 3-41 Sweeping.

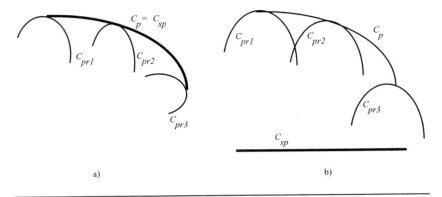

Figure 3-42   Effect of the spine at the creation of a swept surface.

words the sweep pivot, can be defined as the intersection point between the path and the profile curves.

A spine function can be given to the path curve. Figure 3-42a shows this dual function of a path curve that also controls the shape as a spine. In Figure 3-42b, section curves of the swept surface are parallel because the spine is a horizontal straight line.

Figure 3-42 illustrates swept surfaces where identical profile curves are generated along the path. Scaling transformation can be applied to the generator curve as it is swept along the path curve in the direction x (Figure 3-43a) or y (Figure 3-43b) of the plane of the profile curve. Similarly, rotation transformation can be applied to the generation curve along its route (Figure 3-43c). Scaling and rotation along the path can be added linearly or according to a curve. Coordinates x, y, and the normal n constitute a local coordinate system. This coordinate system is moving along the path curve during generation of the surface.

3.2.3.4 Constraining Surface Models

Constraints *leave freedom* for the designer to make subsequent decisions *while protecting the model against breaking earlier decisions*. Constraints are placed by engineers, during interactive sessions, or generated by modeling procedures. Engineers can select, among others, the type of entity and place dimensions, and specify tolerances as constraints.

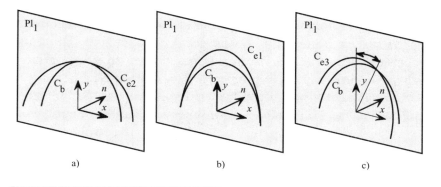

a)   b)   c)

Figure 3-43   Scaling and rotating the generation curve.

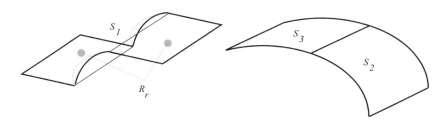

Figure 3-44   Surface related constraints.

Procedures generate constraints to create an entity according to input information, or to maintain specified end surface conditions. The procedure for the creation of the selected entity has built-in knowledge for constraining.

In Figure 3-44, S_3 is constrained as a rotational surface. This constraint is placed by the model-creation procedure. It prevents modification of the analytic shape of the rotational surface at development and application of the model. When the new surface S_2 is integrated in the model, end surface continuity conditions at the connection of surfaces S_2 and S_3 can be established and maintained only by the modification of S_2.

Modeling procedures manage constraints to maintain advanced shapes. Some part designs need local bulges or inflections in their shape. The modeling procedure maintains global smoothness of the surface in the case of these local modifications. Harmonic coexistence of strictly constrained analytic and free form regions can be ensured by *rigidity control*. This allows modification of some regions of a surface while keeping the selected flat, cylindrical, etc., surfaces and their relationships unaltered. In Figure 3-44, the region R_r, consisting of two parallel flat surfaces, is under the protection of rigidity control.

The value of a dimension can be placed in the model in several ways:

By an engineer, during interactive or other direct definition.

By a procedure, as result of a specified or selected calculation.

By a procedure, using associativity defined between the actual and another dimension.

By a shape optimization procedure that uses results from finite element or another analysis.

Engineers modify free form curves and surfaces interactively. *Connector constraints* join curves into more complex curves. Modification of a curve or a connecting point causes modification of all attached curves to save the connector constraints. This action also modifies the surfaces bounded by the modified curves.

A model is fully constrained where modeling procedures generate all shape and dimension definitions as constraints. This shape is considered to be a result of a design and it cannot be modified. A model that contains shapes and dimensions free for modification is under-constrained. In many cases, under-constraining is allowed.

Manufacturing related constraints often force modification of concepts for design related constraining of surface models. Design and manufacturing constraints are defined during concurrent engineering sessions simultaneously. Most manufacturing constraints come from the restricted capabilities of surface machining processes. As a typical example, the die angle constraint specifies a minimum value of the draft angle throughout a surface for the casting or forging of tools.

3.2.3.5 Modeling Complex Surfaces by a Curve Network

This text has reached the point where basic concepts for the creation of complex surfaces are almost complete. An additional advanced concept should be discussed. It is creating a complex surface using a *curve network* as input information. A curve network involves all initial information to create a set of connected surfaces constituting a complex surface for cars, ships, airplanes, and other products with well-engineered complex surfaces. It is an intersecting mesh of curves and includes information about end surface constraints to establish the specified level of continuity across surface boundaries. Its main advantage is that continuity specifications for subsequent surface creations are defined in advance; modification of existing surfaces for this purpose is completely eliminated. Substantial design intent is captured and the overall quality of a complex surface is ensured accordingly. Saving component surface information allows modification of the resultant surface complex as a single surface by control curves.

Entities in a curve network (Figure 3-45) are the curve (C_1), vertex (v_1), edge (e_1), and boundary (B_1). Vertex and edge entities act as topological entities. The intersection points of curves are mapped to vertices. Part of a curve between two intersection

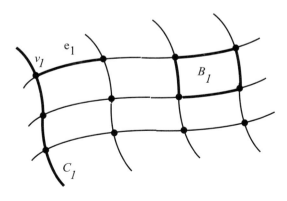

Figure 3-45   Entities in a curve network.

points is mapped to an edge. A curve (C_l) is used to construct several surfaces.

A curve network is processed by the automatic generation of edge curves for individual surfaces. An edge curve is a section of a curve as mapped to an edge. Closed regions are enclosed by closed chains of edge curves, acting as boundary curves. Then all regions are fitted with individual surfaces in a single automatic modeling session. Network surface types are three-sided (Figure 3-46a) or four-sided (Figure 3-46b). Possible T-junctions in the network are collapsed to boundaries for three- and four-sided surfaces (Figure 3-46c).

Closed boundaries of surfaces are automatically defined using topology and curve information from the curve network model. Surfaces are generated then trimmed by closed boundaries. Standard NURBS surfaces are applied for this purpose. The

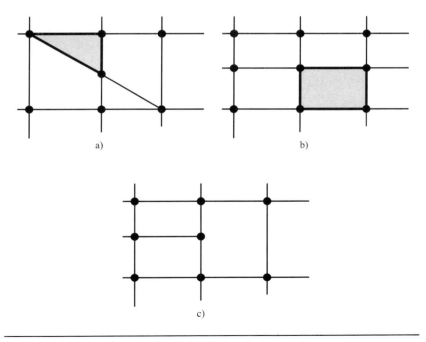

Figure 3-46   Creating surfaces from a curve network.

surface creation procedure maintains the specified degree of continuity along curves in the network. Continuity is enforced on all adjacent surface pairs across the curve. In certain regions of the network where more smoothness is required, a higher degree of continuity can be specified. Parameterization is matched throughout the surface network, along all vertical and horizontal sections.

The overall shape and character of the network driven complex surface can be modified by a *sculpt curve*. A sculpt curve is placed outside of the curve network and projected onto the surfaces in the direction of their normal. Its discrete points are tied to the surface at appropriate projection points with parameters u,v. The influence of a sculpt curve on the surface is defined by control parameters such as *weight* and *region of influence*. Weight is the amount of influence. Increased weight results in projection points approaching the points on the sculpt curve more closely. The weight value may be constant for all points along the length of the sculpt curve, or a weight profile can be defined along the sculpt curve by weight values at selected points. These points are interpolated automatically. The region of influence of a sculpt curve determines the affected area on the surface. It is defined in the parameter space of the surface. This control parameter changes the dimension of the affected region around the projection point.

Representation of ____ Elementary Shapes

This chapter summarizes, completes, and compares elementary curves, elementary surfaces, offset geometric entities, solid primitives, and form features. An elementary shape exists as an individual shape and has its own type, shape characteristics, and attributes. On the other hand, it is a segment or a structural element of a more complex shape and its characteristics and attributes probably depend on other elements in the complex shape.

4.1 Models of Elementary Lines and Curves ____

The choice of elementary shapes starts with the straight line. Straight lines are applied as edges of flat, ruled, or other surfaces. They also serve as important reference entities such as centerlines, etc. Arcs are elements of compound lines, sections for fillets, and serve many other purposes. In addition to lines and arcs, analytical lines are the conics such as circles, ellipses, hyperbolas, and parabolas (Figure 4-1).

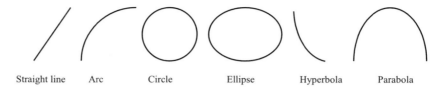

| Straight line | Arc | Circle | Ellipse | Hyperbola | Parabola |

Figure 4-1   Analytical curves.

Deletion of the analytical shape constraint from a rational B-spline curve enables the B-spline to be modified as a free form curve. The effect of a shape constraint can be deleted for the entire curve or only for one or more of its segments. The initial shape of a free form curve is free to change. This freedom is restricted by the representation capabilities of the applied mathematics background. For example, a Bezier curve cannot be modified locally whereas a rational B-spline curve ensures excellent local modification. The shape of a curve is locally restricted by, among others, the shape of the available and economical cutting tools.

4.2   Models of Elementary Surfaces

Figure 4-2 summarizes types of elementary surfaces for industrial design engineering as they were characterized above in Chapter 3.

Fillet surfaces are created along a common edge of two surfaces or between two surfaces having no common edge. A fillet surface is created as an individual surface entity. The model saves information about the original and the fillet surface in the history of the model construction. The radius of a fillet can be:

constant (Figure 4-3b),
linearly variable (Figure 4-3c),
variable according to radius values at given points, or
variable according to a law defined by a curve.

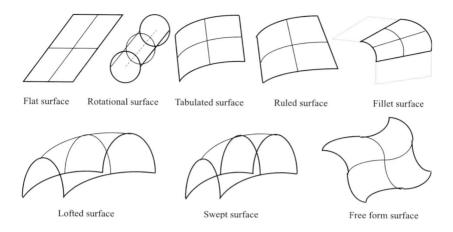

Flat surface   Rotational surface   Tabulated surface   Ruled surface   Fillet surface

Lofted surface   Swept surface   Free form surface

Figure 4-2   Elementary surfaces.

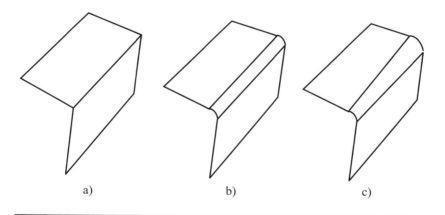

a)   b)   c)

Figure 4-3   Fillet surface.

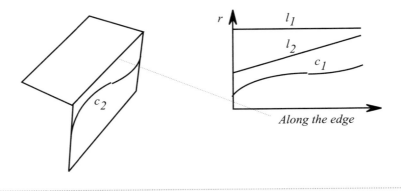

Figure 4-4   Control of the shape of a fillet surface.

Figure 4-4 summarizes the means for control of the fillet radius by lines and curves defined for this purpose. In the plot l_1 is for constant radius, l_2 is for linearly variable radius, and C_1 is for variable radius according to a curve.

The fillet surface in the direction of the edge can also be controlled by a spine. Limiting curves can be defined for the fillet at the connecting surfaces. In Figure 4-4, limiting curve C_2 is defined instead of an automatically created limit. Note that geometric entity definition for control of the fillet shape can produce a geometric processing task that has no solution or cannot be solved by the available modeling procedures. The error message or change proposal from modeling procedures should be followed by a change of input conditions by the engineer.

As shown above, the fillet surface is defined by use of existing model entities. Entities used for this purpose are outlined in Figure 4-5. Simple fillet surface S_f connects surfaces S_1 and S_2 along L_1 (Figure 4-5a). In Figure 4-5b, the definition of a fillet surface on corner point P_1 is propagated automatically to the edges running into it: surfaces S_7, S_8, and S_9 are affected. The reverse task is produced when edges with existing fillets run into P_1. In this case, the corner fillet is controlled by three existing fillets. The geometric conditions can be very complicated when the edges have different fillets. Surfaces S_{14} and S_{15} in Figure 4-5c are to

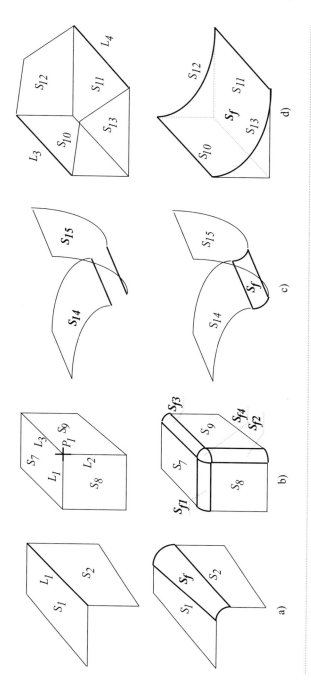

Figure 4-5 Entities to be connected by fillet surfaces.

be connected by a fillet surface. This fillet surface can be defined in the space by its limiting edges L_3 and L_4 and limiting surfaces S_{10}, S_{11}, S_{12}, and S_{13} (Figure 4-5d).

Fillet surfaces are created along open and closed complex contours. The segment of a complex fillet between two fillets with different radii is called a *transitional fillet*. Sometimes a high number of surfaces connected by filleted edges are affected by filleting. This produces many problematic situations for geometric calculations.

A series of filleting operations may produce a complex set of analytic surfaces and curves. Advanced filleting methods can fit a single or several *free form surfaces* to replace a large number of analytical curves and surfaces. Fillets in advanced shape models are often created as free form surfaces instead of as simple analytic surface segments. However, continuity constraints do not allow too many variations of their shape.

The fillet creation process modifies the topological-geometrical structure of boundary representation. A change of geometry often needs a change of topology. Some entities are deleted and new entities are created. A fillet surface and its borderlines are generated, then the surface is trimmed by the borderlines. New topological entities are created to accommodate the newly created curves and surface.

The steps of a basic *fillet creation process* are illustrated by a simple example in Figure 4-6. A fillet is to be created between flat surfaces S_1 and S_2 along line L_1 with radii r_1 and r_2 at the start and end of the fillet, respectively (Figure 4-6a). Fillet surface creation ensures continuity at the connections of the fillet surface to the original surfaces (Figure 4-6e). Fillet surface creation is easy to model by the well-known rolling ball method. It is supposed that a ball with radius of the fillet moves along the edge to be filleted. In the example of Figure 4-6b, the radius of the ball changes between r_1 of the sphere sp_1 and r_2 of the sphere sp_2. The center of the sphere moves along line L_2. Fillet surface boundaries are created on the original surfaces by projection points of L_2 to them (Figure 4-6c). Fillet surface S_3 is created, and then trimmed by

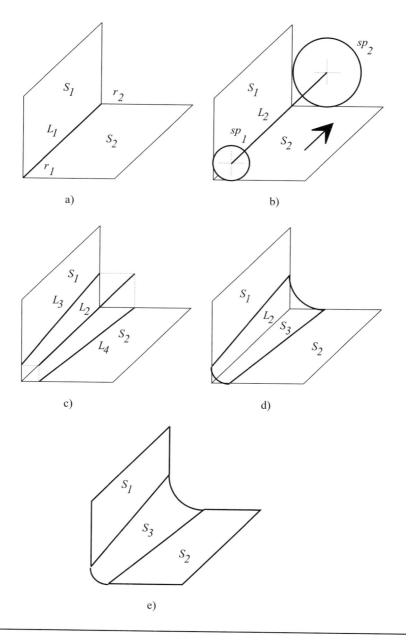

Figure 4-6 Creating a fillet surface in a topological-geometrical structure.

the boundary lines L_3 and L_4 as well as by the arcs at the start and end of the track of the ball (Figure 4-6d). The topological structure is completed by the face for the fillet surface, and four edges for the four boundary curves and vertices.

When the original surfaces are curved, predetermined points along the track of the ball center are projected to the original surfaces. The projected points are interpolated by curve segments. Interpolation does not guarantee that all points of a segment lie in the original surface: a tolerance is specified according to the task. Segments are pulled onto the surfaces as required. Tangent and curvature data along the boundary curves are applied in creating a fillet surface with the specified continuity. The process is the same when the track of the sphere is curved.

Entities resulting from filleting depend on the application of the fillet surface in the construction of parts. The original and fillet surfaces can be placed in different boundaries on different parts (Figure 4-7). This is why the creation of separate fillet surfaces is so important for engineering practice.

4.3   Offset Geometric Entities

Construction and application of part models often require making copies of curve and surface entities at a predetermined distance from the original one. Each point of the copy is at the same distance from the same point on the original. Referring to the method of creation, copies are called *offset curves and surfaces*. Original and offset entities are associative: when the original curve or surface changes, its offset changes accordingly. The offset entity is constrained against any direct change. Direct change of an offset can be allowed by deleting this constraint. This method is important when a shape modeling action is advantageous to start from an offset, as illustrated in Figure 4-8. Modified curves and surfaces are no longer mapped as offsets of their originals.

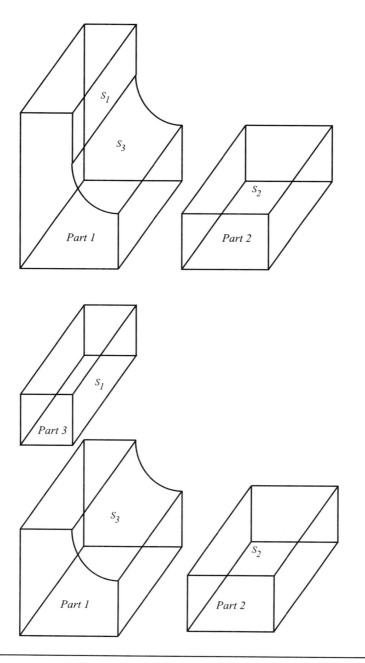

Figure 4-7   Resultant entities of filleting.

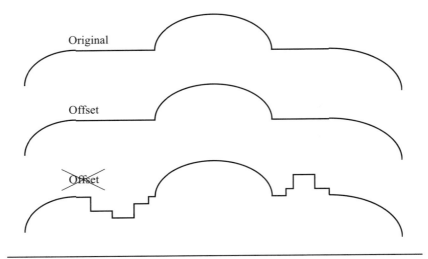

Figure 4-8 Construction of an open contour starting from an offset.

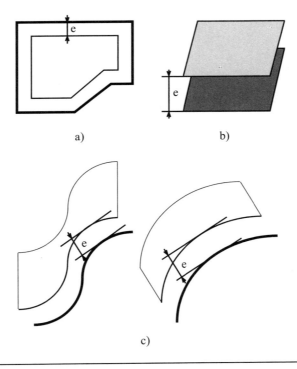

Figure 4-9 Offset curves and surfaces.

Offsetting is a very useful function for everyday modeling practice: Figure 4-9 shows several typical applications. The distance between the original entities and their offsets is *e*. Application of an offset surface is useful for the modeling of hollow bodies with the same outer and inner shape (Figure 4-9a). Original and offset curves are applied to create shapes with walls of identical thickness. Offsets of flat surfaces are applied as widely as offsets of curved surfaces. An inside contour is handled as an offset of an outside contour and vice versa. Offsets of curves such as arcs and free form curves are used as input information for creating new surfaces (Figure 4-9c).

Some geometric configurations make *offsetting not feasible.* The result is not valid where an offset is intersected or looped, as explained by Figure 4-10a. The phenomena of intersecting and looping limit the range of original shapes and offset distances suitable for offsetting. The deviation of the offset curve C_{cr} from the theoretical offset curve C_{th} is an error of offsetting (Figure 4-10b). It is measured at a given parameter value ($u = u_l$) and analyzed where the accuracy of the offset curve or surface is important. The allowable deviation or tolerance range is specified by upper (C_{ul}) and lower (C_{ll}) limit curves. Note that multiple offsetting gives different results than the same value of offsetting in a single step because *deviations are cumulative.*

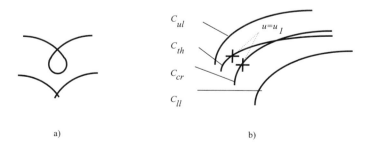

a)   b)

Figure 4-10   Errors in offsetting.

4.4   Elementary Solids

The concept of boundary representation supports both boundary and body centered creation and application of the models. While calculation of cutting-tool paths is generally boundary centered, part design is inherently body centered. Solid topological entities, such as body and lump,[1] are connected with geometric entities through face and edge entities in the boundary representation of parts. The building elements for part models are solid primitives and form features.

4.4.1   Solid Primitives

A solid primitive is prepared for combination with other solid primitives or a more complex solid model under construction. It is created in its final position or repositioned after creation somewhere in the model space. Values of its dimensions are set and the solid primitive is ready for one of the element combination operations. The shapes of primitives are predefined for the modeling system or defined by engineers at application of the modeling system. Users apply one of the available solid generation rules starting from contours, sections, and curves as input entities. *Primitives with predefined shape* are called canonical. They are the cuboid, wedge, cylinder, cone, sphere, and torus (Figure 4-11). Inclusion of shapes other than canonical as predefined shapes is rare because application oriented shape definitions are better to define as form features.

User-defined primitives (Figure 4-12) are created using generation rules that are applied for the generation of elementary surface models. A *tabulated solid primitive* or prism is created by translation of a flat contour along a vector. Rotation of a contour around a centerline produces a *revolved primitive*. Application of an angle

---

[1]See Figure 3-18.

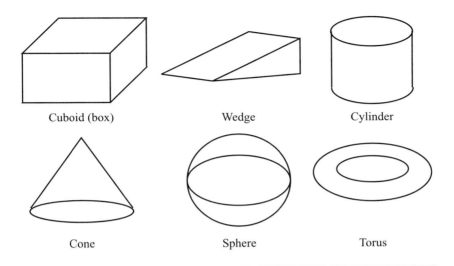

Figure 4-11 Canonical solid primitives.

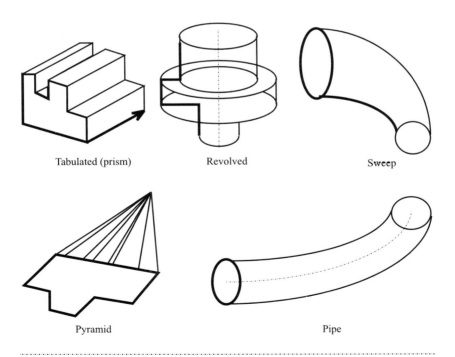

Figure 4-12 User-defined solid primitives.

of rotation of less than 360° results in a revolved solid segment. A *swept solid* is generated as explained for methods for control of shape.[2] Some solid modeling procedures accept an *open generation curve* and automatically close the resultant surface with an additional surface into a solid. A pyramid is created using a contour and a point. Created by moving a circle along an open or closed contour, a *pipe* can be considered as a special sweep where the same contour acts as path, spine, and centerline.

4.4.2 Form Feature Concept in STEP FFIM

The basic concept of modification of a shape by form features is *volume adding and removing*. The shape is then adjusted by fillet and other *treatment features*. While solid primitives are geometric oriented shapes, form features can be application orientated. In other words, form features can be defined according to their purpose and function in the modeled part and use in a part model. At the same time, form features are often defined as pure geometry.

Modeling by form features requires three different sets of information about the application of the modeled shape object, the shape itself, and the representation of the shape. The Form Feature Information Model (FFIM) of the STEP product model standard of the ISO describes the above sets of information on *three levels*.

It was recognized that instances of the same shape carry different information on different parts for different products. For that reason, on *level one* the application information of the shape is described in the form of attributes. This description is called an *application feature*. The form feature is described on *level two* as a *general-purpose shape aspect*. The shape of a part can be defined by different sets of form features from different aspects of the shape. The form feature is defined according to its shape modification

---

[2]See Figures 3-41 to 3-43.

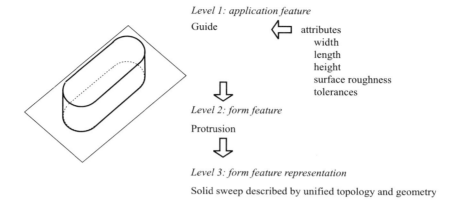

Level 1: application feature

Guide ⟸ attributes
                    width
                    length
                    height
                    surface roughness
                    tolerances
⇩

Level 2: form feature

Protrusion
⇩

Level 3: form feature representation

Solid sweep described by unified topology and geometry

Figure 4-13   Three-level approach for form feature definition.

effect. On *level three* the geometric model representation of the form feature is described. In present practice, a *boundary representation* is attached to each form feature because it supports the description of solids, shape modifications, curves, and surfaces.

The three-level feature model is explained in Figure 4-13 by an example. The application feature "Guide" describes information about dimensions and non-geometric specifications. The form feature defines the shape as a volume-adding *protrusion*. The feature is represented by *unified topology and geometry*, created by *sweeping* as a shape control law.

Basic groups of form features in the FFIM are explained by the example of a rotational part in Figure 4-14. Protrusions *add* while depressions *subtract* volumes. A *transition* may be *flat* as a chamfer or *circular* as a fillet. Volumes can be connected by a *connection volume-adding feature. Through holes* are considered as special volume-subtracting features. The *separated volume* and *void* are volume-adding and volume-subtracting form features, respectively.

The form feature representation records information about the means of shape control. Sweeping is often preferred. Figure 4-15 shows three applications of axisymmetric sweep for creating

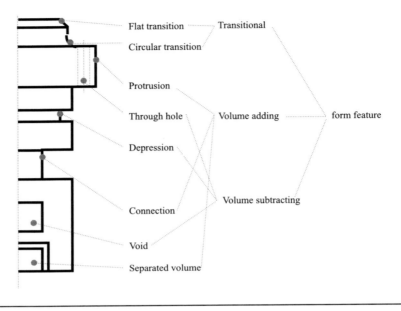

Figure 4-14 Groups of form features in the STEP FFIM.

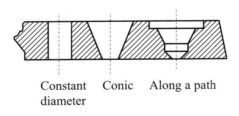

Constant   Conic   Along a path
diameter

Figure 4-15 Application of axisymmetric sweep for definition of form features.

holes by using typical input curves. This is an alternative to rotation around an axis.

4.4.3 Reference Elements for Construction of Form Features

The purpose of *reference elements* is not a shape modification but the assistance of the construction of feature based part models.

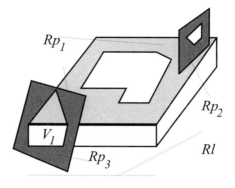

Figure 4-16   Reference elements.

They are simple geometric entities such as points, lines, and planes and relate form features to the modified sections (Figure 4-16). For given steps of model construction, reference elements can be selected from existing entities in a part model under construction. When appropriate reference elements are not available in the part model, they must be constructed and then placed in the model as an entity that is included in the model but is not included in the shape of the part.

Construction plane Rp_1 (Figure 4-16) is selected to define the location of creating a contour based feature. The contour is sketched in its final location by the method "sketch in place." Construction by this method guarantees that all points of the contour lie in the selected plane. Rp_2 is defined to create a contour out of the existing shape. Line Rl is defined as a centerline for a subsequent shape modification. Rp_3 acts as a split plane for the definition of the subtraction of volume V_1. A reference element may be created to control a shape such as a spine at sweeping.

Similarly to coordinate systems, global and local reference planes and lines can be defined. A global reference element can be used for construction of any feature in the shape model, while a local reference plane is used for the feature being constructed.

4.4.4   Types of Form Features

Various implementations of shape definition rules have resulted in comprehensive ranges of form features. Modeling systems offer choices tailored to fields of applications. Engineers define and store form features for groups of products or even for individual products because it is impossible to foresee all the features necessary in the practices of an organization. Most shape modification activities consist of three main steps as follows:

Selection of a plane as a reference entity for modification.
Construction of a flat contour in the reference plane.
Creation of the modifying shape.

Contours are associative with form features. Figure 4-17 illustrates possibilities for shape definition by typical examples for contour based form features. The rule of tabulation of the contour along a line can be considered as the basic method for the creation of volume features (Figure 4-17a). Volume adding or subtracting is done according to the direction of shape modification. Tabulated form features also can be created from an open contour. As a second example, lofted features are created through predefined sections (Figure 4-17b). Sections are often defined in both selected and created planes for the same feature (Figure 4-18). Existing lines and curves can be selected as sections or as elements of compound sections. Some modeling systems allow closed loop features through predefined sections. As a third example, generation curves are swept along a path curve resulting in swept features (Figure 4-17c). Sweep examples illustrate the versatility of the method.

Rotational features can be created using closed or open contours. In Figure 4-18, reference plane Rp_1 is defined outside of the shape for the creation of the closed contour to be rotated; the axis of rotation is Rl_c. Reference plane Rp_2 represents the plane where the shape is modified by the rotational feature.

Rib is a frequently applied element in mechanical parts. A rib feature is created using a contour in a reference plane outside of

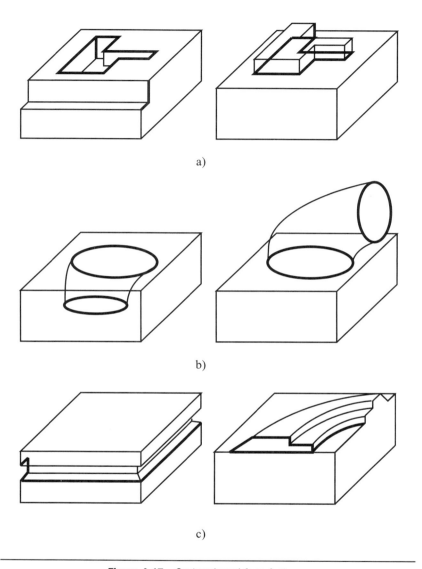

Figure 4-17   Contour based form features.

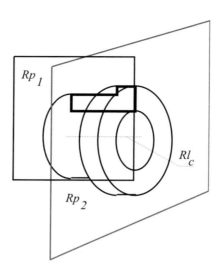

Figure 4-18   Rotational feature.

the existing solid, as explained in Figure 4-19a. Reference plane C_p then contour C are defined. An additional input element is the direction vector v: the width of the rib equals the length of this vector. Surfaces S_1 and S_2 are geometric entities that are mapped to a pair of topological faces to be opened[3] to accommodate the extension of the boundary by the rib. A *network of ribs* can be placed on a part as a single form feature. Figure 4-19b shows the application a network of lines and curves L_n for the creation of a network of ribs. Elementary ribs in the network may have their own width and length values.

Separate surface models can be integrated in the boundary of solids by application of one of the *surface related features*. Topology and geometry are modified to extend the boundary by the surface. Two methods are applied according to the relative position of the surface and the existing solid. Application of the first method results in a closed volume between the surface and the

---

[3]See Figure 3-23.

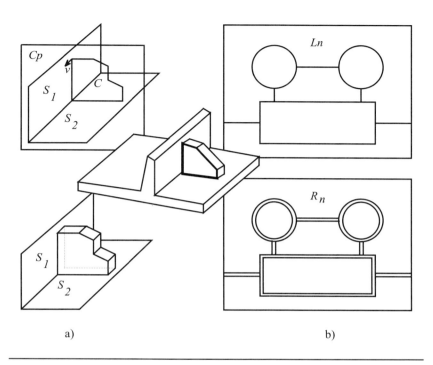

Figure 4-19 Ribs and rib networks.

other surface or a surface of the modified shape (Figure 4-20). The second method applies the intersection of the surface with the modified shape (Figure 4-21).

In Figure 4-20, surface S_1 is used for the creation of different form features. A solid can be created between the surface and its offset (Figure 4-20a). A shape can be closed by projecting the surface onto a planar surface (Figure 4-20b). The feature is defined between the surface and its projection. The surface can be used for shape modification by projecting it onto a planar surface in the boundary of the modified shape (Figure 4-20c).

The intersection method of surface integration assumes that intersection of the surface and the modified shape is geometrically feasible. The surface is placed in its final position relative to the shape to be modified (Figure 4-21). Following this, surfaces in the

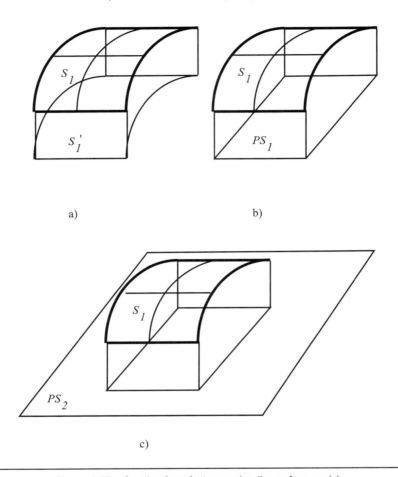

a)   b)

c)

Figure 4-20   Creating form features using the surface model.

boundary of the solid to be modified are intersected by the surface. New surfaces and their trimming curves are created. The topological structure is completed by new edges and faces to allow mapping of new curves and surfaces in the boundary. In Figure 4-21b, multiple sections of a solid are intersected by a surface.

Existing form features are modified by *conditioning* or *dress-up features*. The form feature on which a dress-up feature is defined is called a *support feature*. Dress-up features are *draft, fillet,*

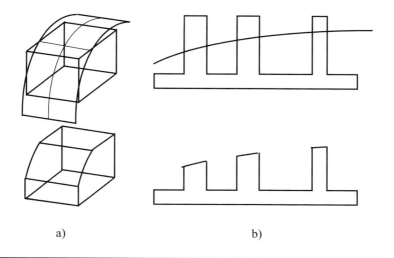

Figure 4-21 Modification of the form-feature based solid model by the surface model.

chamfer, *shell*, *pattern*, and *mirroring*. They are placed on a single support feature or on a branch of support features (Figure 4-22). Conditioning features are contextual with support features by associativities.

The opposite surfaces of prismatic parts, such as of the simple box in Figure 4-23, are generally parallel and the surfaces connected by a common line are at right angles. These surfaces are appropriate when the manufacturing process of the part is machining. Certain manufacturing processes, such as casting and forging, require an angle of a few degrees at surfaces; this is called the draft angle. Other parts contain surfaces tilted by a few degrees. Construction of parts with tilt surfaces is time consuming. Another drawback is that any change or deletion of angles requires repeated reconstruction of the part. Easy definition, change, and deletion of angles are ensured by the application of the conditioning form feature *draft* on tilt surfaces.

The draft form feature describes information about the face to be modified (F_d), the neutral edge that remains in its original position (E_n), the direction of drafting (p), and the draft angle

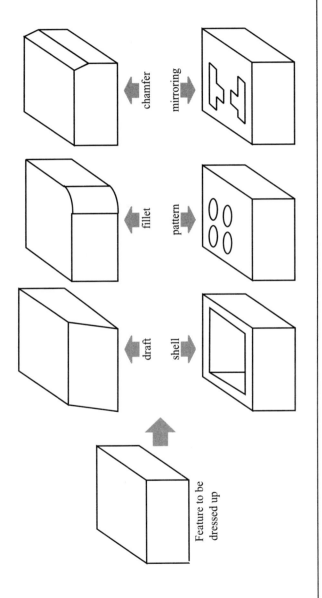

Figure 4-22   Dress-up features.

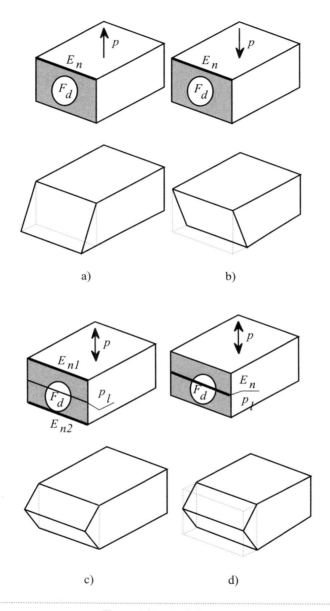

Figure 4-23   Draft features.

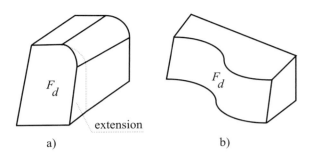

a)   b)

Figure 4-24   Draft on curved features.

(Figure 4-23). If a tool for the manufacture of a part consists of two halves then dual draft is needed (Figures 4-23c and d). Sometimes both directions should be placed on the same surface. For this purpose, draft form feature information is completed by a dividing line (p_l). In Figures 4-23a and c draft form features add volume, in Figures 4-23b and d form features subtract volume. Original shapes are indicated by dashed lines. In the case of Figure 4-23d, the neutral and dividing lines coincide.

When a draft is placed on a curved surface, special geometric conditions may be handled. The draft on face F_d in Figure 4-24a requires extension of a plane and a curved surface. In the case of a draft on a curved surface, as F_d in Figure 4-24b, the draft is created by modification to ensure the specified draft angle on the whole surface. Note that it can be misleading that most of the surface shows the draft angle larger than specified. Critical zones with draft angles smaller than specified cannot be recognized on the screen. Modification of a surface by draft may require breaking of the shape constraint.

Multiple filleting needs coordination of component fillets. Figure 4-25 shows several common combinations for multiple fillets. The *transition fillet* is created to connect two fillets of different radii (Figure 4-25a). A *corner fillet* is defined by filleting a vertex together with the edges running in the vertex (Figure 4-25b). This feature is an alternative to the creation of three fillets

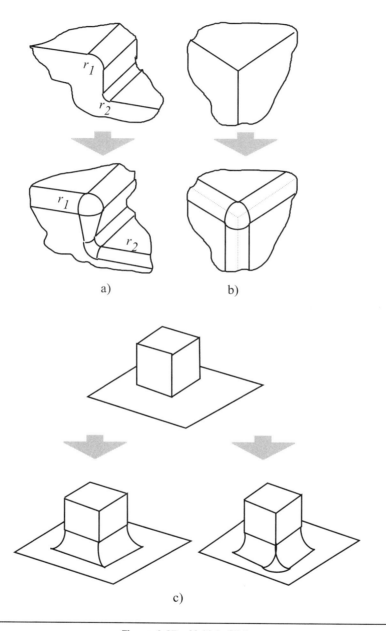

a)   b)

c)

Figure 4-25   Multiple fillets.

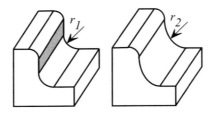

Figure 4-26 Fillets.

individually. Two basic configurations of fillets around a rectangle at the connection of a cube and a plane surface are created with sharp and round corners (Figure 4-25c).

A change of fillet geometry may change the topology as in the case of Figure 4-26 where the face mapped to the shaded surface, three edges, and two vertices are removed as a consequence of a changed fillet radius from r_1 to r_2.

Hollow parts with the same outside and inside shape, with uniform or wall-dependent thickness, need volume subtraction-type shape modification to create an inner boundary. These parts are defined by the outside shape and wall thickness instead of the outside shape and depression. At application of depression as a form feature for inner regions, a change of wall thickness would require redefinition of the depression feature. Shelling shape modification procedures ask for information about the surfaces to be removed to open the boundary and wall thickness values; then they create inside areas with the specified wall thickness. Four examples in Figure 4-27 introduce the modeling capabilities of this dress-up feature. The boundary of the part is opened at one, two, three, and four surfaces. In the case of four opened surfaces, the original solid is split into two separate lumps.

Multiple occurrence of a feature is defined by placing child features of a parent feature along a specified pattern (Figures 4-28a, b) or by mirroring (Figure 4-28c). Children are associative with their parent. Any change of the parent results in the

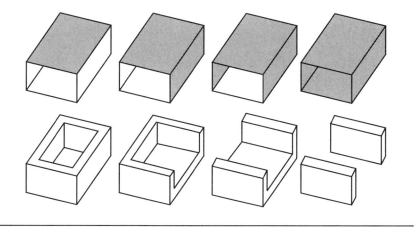

Figure 4-27 Shellings.

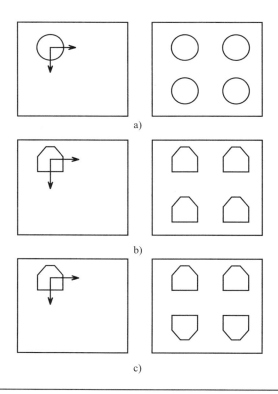

a)

b)

c)

Figure 4-28 Patterns and mirrorings.

same change of its children except for individually configured children.

The types of form features discussed above include the important ones although the list cannot be complete. A variety of other feature definitions such as helical protrusions and cutouts, teeth, etc., can be available to modify shapes of parts.

Models of
Shape-centered Products

5.1 Models of Combined Shapes

Combined shapes are defined as combinations of elementary shapes utilizing the expertise of engineers in part design and shape modeling. Combined shapes are created by the application of one of the following methods:

> *connecting elementary surfaces* using intersection and blending;
>
> creating groups of surfaces from *curve network* information;
>
> *combination of solid primitives* using constructive solid geometry;
>
> sequence *of shape modifications* using form features.

Creation of combined shapes is assisted by *control techniques* built into the modeling procedures, such as consistency analysis, rules, and checks. They help engineers to make correct models and they minimize the risk of errors encountered through engineers who are experts in part design but inexperienced in shape modeling.

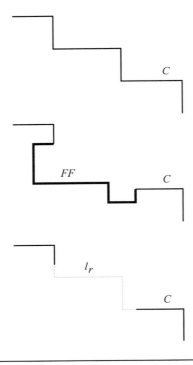

Figure 5-1   Modification and reconstruction of a shape.

Construction of models often requires deleting, modifying, replacing, or only repositioning of entities or their attributes. When these activities are considered to be unsuccessful, one of the earlier stages of the model should be reconstructed. This requires reversible model construction. Another problem is that the deletion of entities on complex or intricate shapes may leave complex shapes to be repaired. On the other hand, a replacement entity may require modification of a complex shaped region.

Reconstruction of a previous shape is illustrated in Figure 5-1[1]. Contour C is a section of a detail in a boundary. Form

---

[1]Examples in Figures 5-1 and 5-2 are illustrated in two dimensions for simple explanation. Methods of model construction in three dimensions are discussed in Chapter 8.

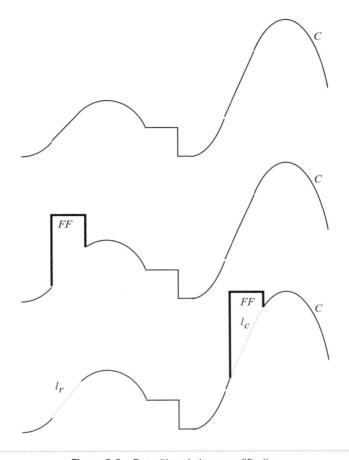

Figure 5-2   Repositioned shape modification.

feature *FF* modifies *C*. At a later stage of construction, *FF* is deleted and the contour is repaired by line chain l_r.

Creation of a new part variant may require repositioning of some existing shape elements. When a shape element is repositioned, regions of the old and new position must be changed and reconstructed, respectively. Contour *C* in Figure 5-2 consists of linear and curved elements. Form feature *FF* modifies it to make a protrusion to accommodate some structural parts. When *FF* is repositioned along *C*, curve l_r is applied to repair *C* and l_c is removed by the repositioned shape *FF*.

5.1.1   History of Model Construction

Are models created with or without a history of construction? This is a question, similar to many others, that can be answered only at the workbench, with the knowledge of the actual modeling task.

Modification of results of earlier steps during and after construction of a model needs information about the affected modeling operations and model entities. Most modeling systems record information about the construction process in the *model construction history*. Modification of a model can be done simply by a step back to the operation where the entity was created, input of the modified parameters, and regeneration of the model according to the modified parameters. This method inherently propagates the effect of modification to the entire model. Note that the feasibility of certain operations may be changed by changed geometry.

Opinions vary about modeling with or without a history. Most recent modeling systems offer construction both with and without a history. The main drawback of the pure history based modification is that any model change is indirect, through a change of the modeling process. On the other hand, history independent modification can break design intent or constraints. Recording the construction history captures modeling methodology and design expertise. Considering the history of construction of an existing model, an earlier model creation procedure can be acquired and utilized for the modeling of new or changed parts by other authorized engineers. The construction history may have an important role in the assurance of model quality. In cases when engineers make history independent modifications during history based modeling, the history may be removed by the modeling procedure.

The description of the structural characteristics of a shape, e.g., sequence of shape modifications or tree of element combinations, is also considered as construction history. History provides access to any modeled object at any step in a design process. Because new variants of a part often can be defined by a minor change of an existing shape, modification of a shape model using history facilitates the design of customized parts for customized products.

A method called *dynamic modeling* represents a trend that abandons any dependence on history. This modeling isolates features but describes the geometry without any history. It accelerates design changes when local geometry and topology changes are needed or arbitrarily placed dimensions are to be defined at any stage of model construction.

Using history information, a model can be changed by *adding*, *removing*, *replacing*, *repositioning*, or *re-attributing* elementary surfaces, solid primitives (Figure 5-3), and form features in a

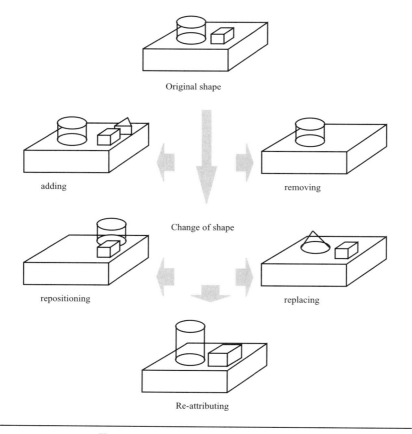

Original shape

adding

Change of shape

removing

repositioning

replacing

Re-attributing

Figure 5-3   Changes of a solid shape model.

combined shape. Temporary or part variant dependent inactivity of elementary shapes can be set by their *suppression*. Information about a suppressed elementary shape remains in the model but this shape does not affect the shape of the part.

5.1.2 Part Models by Constructive Solid Geometry

The shape of a mechanical part can be divided into a well-defined set of solid primitives. A purposeful sequence of combination operations with the primitives can be applied to form the shape of the part. Constructive solid geometry (CSG) is based on this recognition. CSG was the traditional way of solid modeling. The construction method is also applied in advanced part modeling. While the traditional method applied *CSG data structure*, present modeling methods generate *boundary representation*. This difference often causes misunderstanding around CSG.

Figure 5-4 summarizes the procedure of solid shape modeling by CSG. The modeling system offers canonic primitives and procedures for the creation of primitives starting from close contours, vectors, and curves. Solid primitives are selected (Figure 5-4a) and then adapted to the modeling task by the definition of dimensions (Figure 5-4b). Then the primitives are placed at their final positions in the model space (Figure 5-4c). Sometimes primitives are created at their final positions. Following this, pairs of primitives are combined by *union*, *difference*, and *intersection* operations (Figure 5-4d). Operators build the shape. They are called *building*, *combination*, *set*, or *Boolean* operators because they build the shape, make combinations, set primitives, and combination operations in analogy with Boolean algebra.

Construction operations can be represented in a binary tree structure called a CSG tree. The part in Figure 5-5a is created by a sequence of combination operations according to Figure 5-5b. A binary tree (Figure 5-5c) represents the final shape by its root node,

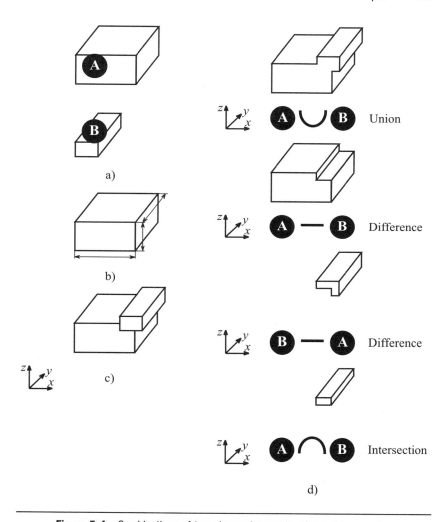

Figure 5-4 Combinations of two shapes by constructive solid geometry.

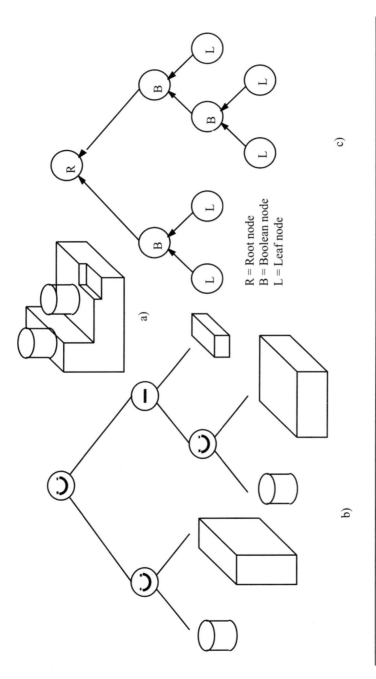

R = Root node
B = Boolean node
L = Leaf node

Figure 5-5   Binary tree for constructive solid modeling.

the combination operations by its Boolean nodes, and the solid primitives by its leaves.

The traditional CSG model data structure consisted of a CSG tree and information about the type, dimensions, and position of primitives. This model was called unevaluated because it did not contain results of intersections. The final shape was generated during application of the model mainly for visualization and geometric calculations. A great advantage of this model was its small demand for memory and storage capacity at that time.

CSG is still offered by most of the form feature based modeling systems because some engineers and engineering tasks demand it as an auxiliary construction tool. Figure 5-6 illustrates a *mixed application of shape modification and element combination in the solid modeling* of a single part and indicates the fundamental difference between the two basic methods of shape construction. Base feature *BF* is modified by form feature *FF* starting from closed contour *CC*. Following this, solid primitive *SP* is created outside of the shape and then transformed to its final position and fused into the solid.

5.1.3 Part Models by Shape Modifications

The text and illustrations in this book have outlined the application of the advanced method of shape modification for the construction of part models. Feature based part modeling describes information about *form features* and the *construction process*. Features may be retrieved from feature libraries, created on site, or defined as modifications or combinations of existing features. Features can be grouped at different levels in order to give structure to the design. The sequence of shape modifications can be reordered when a new sequence is more appropriate for the subsequent steps of construction or modification of the part model, as illustrated by the example of Figure 5-7. Feature suppression makes it possible to inactivate then activate features or groups of features.

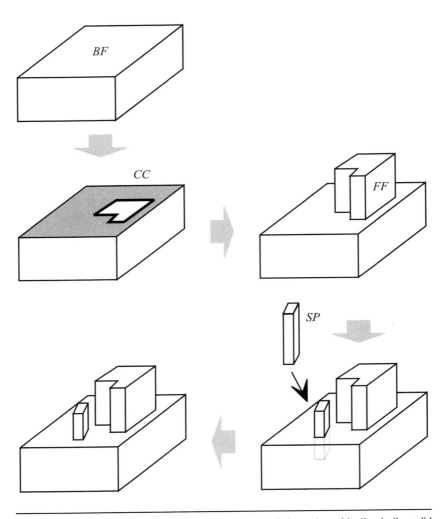

Figure 5-6 Mixed application of shape modification and element combination in the solid modeling of mechanical parts.

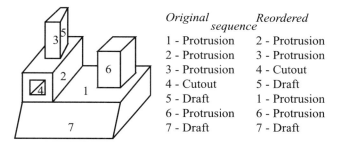

Original sequence	Reordered
1 - Protrusion	2 - Protrusion
2 - Protrusion	3 - Protrusion
3 - Protrusion	4 - Cutout
4 - Cutout	5 - Draft
5 - Draft	1 - Protrusion
6 - Protrusion	6 - Protrusion
7 - Draft	7 - Draft

Figure 5-7   Sequence of shape modifications and its reordering.

5.1.4   Models of Sheet Metal Parts

Sheet metals and the manufacturing processes to form them into sheet metal parts are simple and cheap. Application of sheet metal parts is still very popular. Parts produced by bending and punching processes require special description of the bends and flat patterns in the model.

Models of sheet metal parts are different from any other part models (Figure 5-8). Panels, in other words, tables or faces, are defined and then related by bends. Bend panels are placed in the model space. A bend entity is defined between two panels. Its parameters are thickness, bend type, radius, and allowance. Because the manufacturing of sheet metal parts starts from a sheet, a flat pattern of the part is necessary. Welding or soldering of panels can be coded by a zero value of bend radius. 3D and flat pattern representations are associative so that a flat pattern can be applied to make a 3D representation and vice versa. Changes of each representation are automatically copied to the other one. The engineer can work in both the folded and unfolded state of sheet metal parts. The flattened length differs from the folded length: the difference is the *bend allowance* or *K-factor*. The bend allowance depends on the type of material.

One of the main applications of sheet metal parts is in covering assemblies. In Figure 5-8, an assembly consisting of parts P_{a1}–P_{a3} is covered by sheet metal part P_{sh}. Panels are extracted

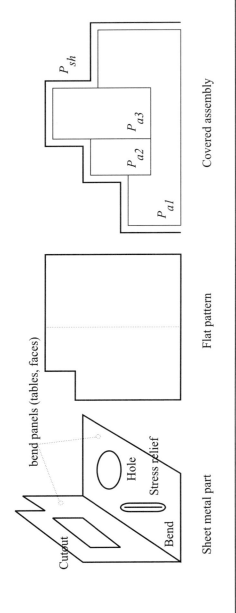

Figure 5-8 Sheet metal part.

from the flat surfaces of covered parts. We say that the sheet metal part is defined within the context of the surrounding assembly. Covered and sheet metal parts can be associative. In this case, dimensional and geometric constraints ensure the correct sheet metal part dimensions. Sheet metal parts can be updated automatically, in accordance with part design changes.

Stress reliefs are defined on the basis of stress analysis. Sheet metal part form features such as punches, tabs, holes, depressions, channels, louvers, etc., can be placed on panels.

Advanced sheet metal part models are represented as solid. Interference between the sheet metal and the covered parts is checked for automatically by use of covered and covering part volumes and assembly model information. When moving parts are covered, interference is checked for critical positions.

5.2 Assembly Models

While advancement in part modeling was enforced by computer controlled machining, assembly drawing remained the main communication medium for assembly processes during the 1970s and 1980s. Assembly drawings, in the form of blueprints or on screen, serve the visual understanding of assemblies. During the 1990s, a much more competitive environment urged the development of assembly modeling. The main difference between the application of part and assembly models is that part models serve the computer control of manufacturing equipment while the assembly model is mainly used as input information for product design.

When the fast development of several computer based engineering activities enforced it, the wide application of assembly modeling was started in the following areas:

Construction of part models in the context of assembly.
Assembly model information for the modeling of kinematics.
Quick evaluation of the effects of design changes in an assembly.

Quick creation of part and assembly variants for product variants.

Quick checking of assemblies for interference of parts.

Producing bill of materials (BOM) information for materials requirements planning (MRP).

Calculation of mass properties of assemblies.

Model based presentations and animations.

An assembly model can be considered as an advanced one when it is suitable for the description of all the information demanded by the above listed activities. The part oriented approach to the modeling of mechanical systems has been replaced by a more comprehensive and integrated approach that includes the modeling of assemblies and kinematics. In fact, assembly modeling has developed into a key position in concurrent associative modeling.

When a part is placed in an assembly, it is an *instance*. Other instances of the same part may be placed in the same or other assemblies. The part model is normally stored as a single copy, it is not duplicated in the assembly model for the instances. The part instance includes information about its orientation and position in the assembly (Figure 5-9). Part modeling is done in associative model spaces of the assembly and part. The engineer can change the space for efficient and comfortable model construction. Some form features on a part depend on connection with other parts, others are the same on different parts. They are easier to define in the assembly space. Assembly-independent details of part models are constructed in model spaces of parts. Part instances have new orientations in the assembly space according to their associativities with other parts in the assembly. Part orientation in the part model space is independent of the assembly; it may be set as comfortable for part modeling. In associative modeling, any change of the part model initiates the necessary changes of all the assembly models where that part is mapped as an instance.

The following applications of assembly models have primary importance.

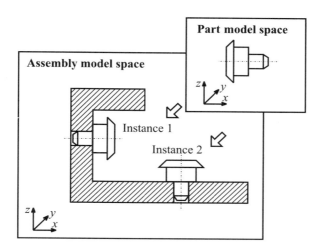

Figure 5-9   Part instances in an assembly.

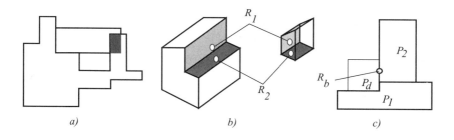

Figure 5-10   Analyses of assembly model.

Collision analysis: Parts often have complicated and intricate shapes. When a part is placed in an assembly, its regions can interfere with regions of previously placed parts. In Figure 5-10a, the dark section represents an area that is occupied by two parts. The solution is an appropriate change of one or both of the parts. Collision analysis is especially important when a new assembly variant is created by quick changes of several parts.

Analysis of consistency: Complex networks of assembly relationships are not easy to understand by engineers during the design and modification of assemblies. Incorrect definition and modification of assemblies often produce inconsistent sets of relationships. When an engineer considers an assembly as finished, a consistency analysis is recommended to reveal any parts or relationship definitions still required. This analysis helps to avoid incomplete or redundant sets of relationships. At the same time, assembly modeling systems offer in-process analysis for incomplete or broken constraints during assembly design. In Figure 5-10b, assembly relationships R_1 and R_2 constitute an under-constrained or in other words not fully placed part. Removing a part, as illustrated in Figure 5-10c, may remove a reference element of one or more relationship definitions. Deleting part P_d removes one of the contacting surfaces for relationship R_b. This situation is called a *broken relationship*. The broken relationship or removed reference element should be replaced.

Creating an assembly drawing and exploded view: A drawing and exploded view are created and modified associative automatically with the assembly model.

Creating an assembly plan: The assembly model is checked for disassembly and a verified disassembly sequence is created. Following this the assembly sequence is defined as the reversed disassembly sequence.

Creating a bill of materials: In products for shape-centered modeling systems, the demand for most raw materials, parts, and assemblies depends on the demand of the finished product. Materials requirements planning (MRP) is an inherently computer assisted method for the calculation of this dependent demand. One of the MRP inputs is the bill of materials (BOM). The bill of materials is a record of information about the identification and quantity of raw materials, components, subassemblies, and assemblies for a product. Standard and other purchased parts, such as fasteners, need special handling. They are neither modeled nor produced in the engineering and production system.

The building of a product by subassemblies and parts is represented in the assembly model structure. The structure is broken into levels. In the example in Figure 5-11, a four-level structure is applied. *Level 0* (L0) is the level of product. *Level 1* (L1) is the level

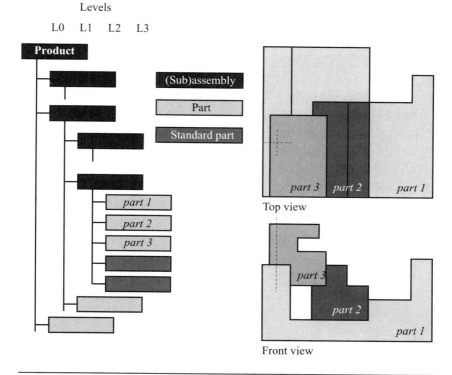

Figure 5-11   Assembly structure.

of subassemblies and parts that are applied for the final assembly of the product. *Level 2* (L2) parts and subassemblies constitute *Level 1* subassemblies. Finally, *Level 3* (L3) parts and subassemblies constitute *Level 2* subassemblies. The number of levels depends on the product.

Purchased assemblies are described as parts in the assembly structure. An engine, for example, is considered as a part in the final assembly of a car. A generator is considered as a part in the assembly of an engine. A ball bearing is considered as a part in the assembly of a generator.

The creation of a bill of materials applies an assembly structure description (Figure 5-12a). The bill of materials describes the quantity per product and source for each part and subassembly.

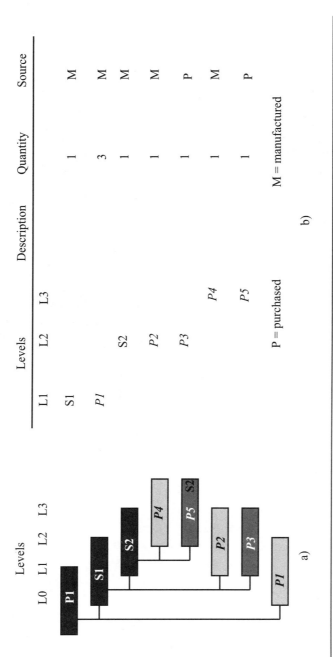

Figure 5-12 Product structure and bill of materials.

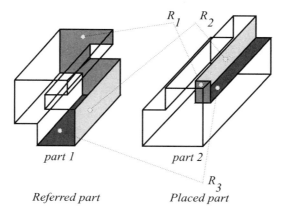

R_1 R_2

part 1   *part 2*

R_3

Referred part   *Placed part*

Figure 5-13 Assembly relationships.

Parts and subassemblies are listed level by level according to the product structure. The source of an item (subassembly or part) may be manufactured or purchased.

Parts are placed in the assembly in relation to other parts by contact, coincidence, and other relationships. Relationships are associativities between parts and act as constraints. They define the assembly and any change of them produces a changed or erroneous assembly. In Figure 5-13, three pairs of flat surfaces are in contact between *part 1* and *part 2*. The result is *complete positioning* of *part 2* by its placing on *part 1*. A complete set of relationship definitions carries *placement information for a part* in the assembly model. A placed part is in connection with one or more referred parts through relationship definitions. Automatic positioning and relationship definition by automatic orientation and translations are offered by advanced procedures.

A complete set of assembly relationships for placing a part does not leave undefined components of its position definition. In Figure 5-14, the relationship R_1 controls one of the components of the part position. The position of the part is still undefined in two other directions. Two additional relationships, R_2 and R_3, are to be defined for full positioning of the part.

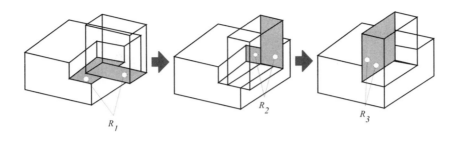

Figure 5-14   Placing a part step-by-step.

Modification of a dimension of a part often requires separation of contacting surfaces and breaking a relationship, as in Figure 5-15a. Associative parts can be moved or re-dimensioned in order to maintain assembly relationships. In Figures 5-15a and b, relationships R_1, R_2, and a third relationship that cannot be seen place *Part 1* in relation to *Part 2*. Placing *Part 3* by a new relationship R_6 and other relationships would break relationship R_2. One of the possible solutions is translation of *Part 2* to the position that allows for saving relationship R_2. This transformation could be done because the original position of *Part 2* was not constrained. An alternative solution is to increase dimension L_1 to L'_1 on *Part 2*.

The parts in Figure 5-16 are placed by, among others, contact assembly relationships R_1, R_2, and R_3. Decrease of dimension H_2 breaks relationship R_2. Fortunately, dimension H_1 is not constrained. The solution is to increase dimension H_1 to such an extent so as to compensate for the decrease of dimension H_2. Analysis and recommended solutions are available in advanced assembly modeling.

Modeling of parts in their own and assembly model spaces is illustrated in Figure 5-17. Models of *Part 1*, *Part 2*, and *Part 3* are constructed individually in their own model space because construction of *Assembly 1* supposes their model. The model of *Part 4* is constructed in the model space of *Assembly 1* because contour C_4 is composed by using contours taken from parts *Part 1*, *Part 2*, and *Part 3*. A tabulated solid is defined in the assembly model

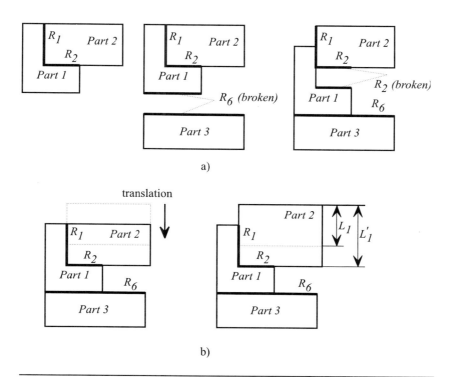

a)

b)

Figure 5-15   Solution for the problem of a broken relationship.

space for *Part 4* and then other details of this part are constructed in its own model space. Assembly form feature *hole H1* goes through parts *Part 1*, *Part 2*, and *Part 3*. It is also created in the assembly model space. Associativity definitions between models of *Assembly 1* and parts *Part 1–Part 3* control the appropriate form features on *Part 1*, *Part 2*, and *Part 3*. *Part 4* and hole *H1* are defined in the context of *Assembly 1*. If this assembly form feature is not available in the modeling system, segments of hole *H1* should be constructed on the three parts.

The assembly model describes a ready-made state of the assembly. When this position is collision-free, feasibility of collision-free assembly and disassembly are the next criteria of successful assembly design. Volumes of other parts in the assembly restrict

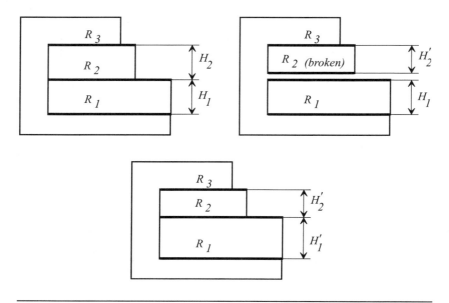

Figure 5-16 Associative relationships.

the assembly and disassembly of a part. Analysis may be applied to find collision-free sequences of paths for removing a part from an assembly at a given stage of construction. Disassembly sequence variants may be considered. If disassembly is feasible, it is supposed that the construction also can be assembled.

5.3 Models of Mechanisms

The modeling and simulation of kinematics ensure the definition and assessment of movements and loads of mechanisms in the virtual world. Input moves, movements of parts, external loads, and movement-generated loads are considered. The model of a mechanism (Figure 5-18) can be considered as an extension of the assembly model with the model of the kinematics. The ability of parts for relative movements is expressed by the degree of freedom in the joint entity. Associativities can be defined between

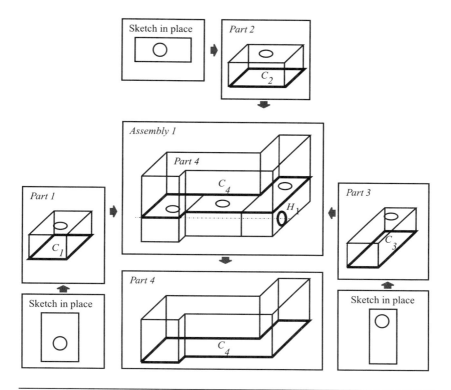

Figure 5-17   Construction of part models in the context of assembly.

joints and part geometry depending on the type of joint. Contact relationships defined in the assembly model may be automatically proposed as potential joints by the modeling procedure during the construction of the kinematics. One of the joints in the mechanism receives input moves from outside. Parts are considered as rigid bodies and modeled as rods. One of the rods in the mechanism has a fixed position. It is called the frame of the mechanism. Joints and rods constitute closed or open loops. The degree of freedom of movement of the whole assembly is determined by superposition of elementary degrees of freedom at joints.

The basic concept of assembly model based modeling of kinematics is explained in Figure 5-19. Parts as rods *R1–R4* are joined by joints *J1–J4*. Figure 5-19a shows the mechanism arranged in a

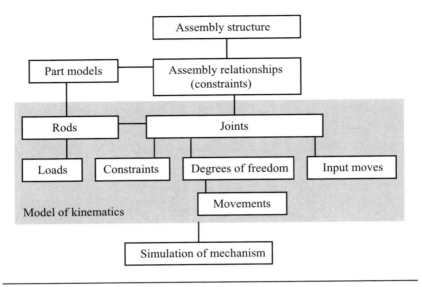

Figure 5-18 Mechanism modeling.

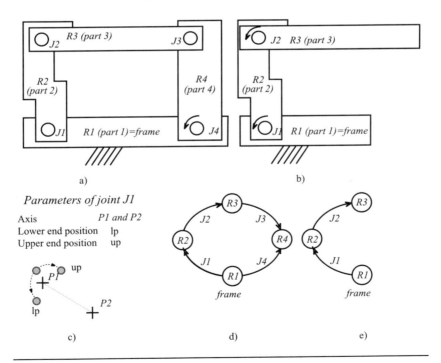

Figure 5-19 Modeling of kinematics.

closed loop. All joints have one rotation type degree of freedom. Parameters of joint *J1*, given as the axis and limits of rotation, are detailed in Figure 5-19c. A rotational input motion is applied at joint *J4*. The question is whether the mechanism can move or not, considering the degrees of freedom at all other joints and the length of the rods. The most important simulation evaluates the degrees of freedom and other limitations for the whole mechanism and answers this question. The structure of the mechanism in Figure 5-19a is represented in the model as can be seen in Figure 5-19d. The mechanism with an open loop is illustrated in Figure 5-19b where two rotational motion inputs result in a robot-like mechanism.

A joint is defined by its type, parameters, and auxiliary entities. The type of joint determines the degrees of freedom. Auxiliary entities are the reference points and lines needed for the definition of movements allowed by the degrees of freedom. Figure 5-20 introduces common joints from the everyday design of mechanisms by their motions, degrees of freedom, and auxiliary entities.

Modeling of curves and surfaces has widened the capability of kinematic modeling. Joints can be defined for movements along and around a curve (Figure 5-21). The rod can slide along the curve, roll on the curve, and rotate around the curve according to the degrees of freedom allowed for the movements. A point-curve type joint also can be defined.

In some mechanisms, parts move along surfaces. Two translations and three rotations may be required to define these motions. Figure 5-22 illustrates two typical combinations.

Finally, several special joints should be mentioned. Cam, gear, rack and pinion, and other components need to be composed of rotational and translational joints. Parts in these joints must follow each other, as is required by the principles of their operation. This is described in the model of the joint by an additional constraint. Relative motion between two translational, rotational, or cylindrical joints and any combination of two of these joints may be predefined. A constant velocity joint serves special applications. A fixed joint has zero degrees of freedom.

Joint type	Motions	Degrees of freedom	Auxiliary entity
Rotational	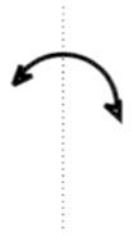	1 rotation	Line (axis)
Translational		1 translation	Line (path)
Cylindrical		1 rotation + 1 translation	Line (axis and path)
Screw		Combined rotation + translation	Line (axis)
Universal		2 rotations	Two lines (axes)
Ball	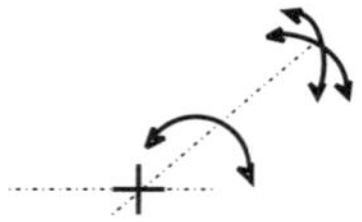	3 rotations	Point (centre)

Figure 5-20 Common joints.

The model of a mechanism as an integrated structure of parts, assemblies, and kinematics has gained its primary application in simulations. Simulations are essential to the understanding of the kinematic and dynamic behaviors of the design under different motion and load circumstances. Mechanism motions are defined using mathematical expressions. Complex functions are used to define complex motions. Time, distance, or angle of rotation can

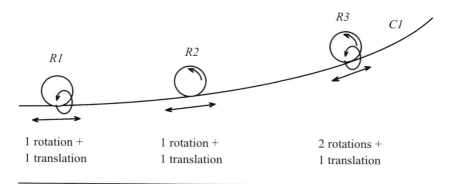

1 rotation +   1 rotation +   2 rotations +
1 translation   1 translation   1 translation

Figure 5-21 Joint definitions using a curve.

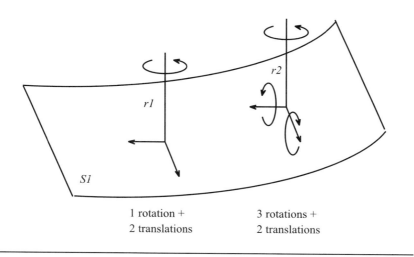

1 rotation +   3 rotations +
2 translations   2 translations

Figure 5-22 Joint definitions using a surface.

be specified to drive a mechanism. Loads must include gravity and friction. Springs and dampers are also considered.

Feasible simulations are determined by the capabilities of the kinematic solver available in the modeling system. Simulation of a mechanism may include the following analyses:

Computing and display motions as a result of given input moves, according to an engineer-defined step rate.

The *envelope* of the volume occupied by a mechanism over time. Shapes of surrounding parts can be derived or modified.

Calculation of position velocity and *acceleration* on any part or between any two parts within the mechanism.

Calculation of resultant forces. Dynamical analysis for joint forces, spring forces, inertia forces, torque, and reaction forces.

Degrees-of-freedom calculation for the entire mechanism on the basis of the degrees of freedom specified for the joints.

Joint validity checking.

Interference and clearance checking, at any path of a mechanism. This analysis also provides information about the integrity of the assembly. Solid models of parts should be available for this purpose.

Ability to follow a path defined by a curve.

5.4   Definition of Dimensions and Tolerances

These areas of modeling deal with the description of the most frequent results of decisions in shape related engineering. Construction of models is dimension driven. Although dimension driven modeling of parts and assemblies was one of the new leading concepts in part modeling during the 1980s, the wide application of fully integrated dimensioning and tolerancing methods is a recent achievement. *Dimension definitions* in a model serve to understand the checking and to improve the dimensioning of a part or assembly by computer analysis. Another application of dimension definitions is the automatic dimensioning of part and assembly drawings using associativities between shape models and engineering drawings.

5.4.1   Definition of Dimensions

Modeling of a part involves two main shape related activities: the definition of geometry and the definition of dimensions. In early

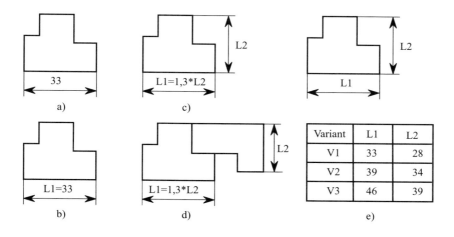

Figure 5-23 Definition of dimensions.

part modeling, dimensions could not be described in the models or only fixed dimension values could be involved (Figure 5-23a). The next stage of development was the definition of dimensions in the form of variables (Figure 5-23b). The description of part variants needed the representation of relationships between pairs of dimensions. The method that uses relationships between dimensions within a part (Figure 5-23c) or within an assembly (Figure 5-23d) is called parametric design. When relationships cannot be defined between the dimensions of a part or an assembly, variant geometry describes lists of dimension values for each variant in the form of a table (Figure 5-23e) or a matrix. In an advanced method, the model of a master part describes the geometry and all associated variables. This model serves as the basis for the generation of a family of parts. A spreadsheet contains alternative values for each of the controlling variables. Relationships between dimensions are described as associativities. Dimension definitions may apply user-written equations. Dimensions should be handled after modifications of shape such as intersection, cutting, and joining.

A dimension definition represents the intent of the engineer who made the related decision. Its content is handled as a constraint and may be created in the form of

a single value,
a range of allowed values,
a logical expression,
a geometric relationship,
a formula or rule to calculate its actual value, or
associativities with other parameters including dimensions
(Figure 5-24).

The geometric relationship may be parallel, perpendicular, tangent, collinear, or coincident. Associativity may define the relationship to the same part, to another part, or to a circumstance in a product, a production, or a product application environment.

Modification of a dimension normally requires modification of other dimensions. They may be calculated using formulas, rules, and associativities in the model. Allowed ranges, formulas, rules, and associativities in a dimension definition prepare *decisions* for the modification of dimension values at subsequent construction and downstream *applications* of the model.

When an associativity is intended to be saved during the modification of related dimensions, it is defined as a constraint. Figure 5-25 shows what effect a constraint has on the consequent modification of a dimension. Length $L1$ is modified to a value within the allowed range in Figures 5-25a and 5-25b. The resultant shape is controlled by additional constraints. In Figure 5-25a, the shape

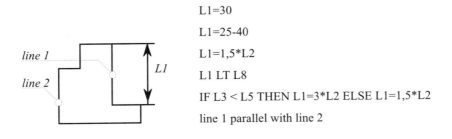

$$L1=30$$
$$L1=25\text{-}40$$
$$L1=1,5*L2$$
$$L1 \; LT \; L8$$
IF $L3 < L5$ THEN $L1=3*L2$ ELSE $L1=1,5*L2$
line 1 parallel with line 2

Figure 5-24   Definition of a dimension.

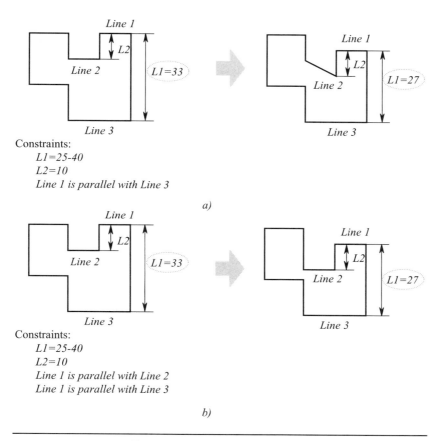

Figure 5-25   Constrained dimensions.

is changed to fulfill the parallel constraint between *Line 1* and *Line 3* and constrained to fix the value of *L2*. An additional parallel constraint between *Line 1* and *Line 2* in Figure 5-25b prevents the part from modification of the shape. Consequently, other non-constrained dimensions must be changed. Note the decrease of the section width at the left end of *Line 2*. To avoid a decrease below a critical value, a minimal distance between *Line 2* and *Line 3* can be defined as a constraint.

The method of dimension and shape definition by constraints is often called variational geometry. It has had dramatic effects on the style of model construction. It is not necessary to create contours for form feature definition by exact dimensions. A sketch is appropriate to put an approximate concept into the computer; then *constraints enforce* the final contour. Note that constraining defines the shape and dimension content to be described in the model instead of a precise drawing on the screen. Associativity and constraint definitions can be included in the form of inheritance of dimensions in object oriented models and they can be specified between parts of different products. Decisions require a given set of constraints. An under-constrained model may lead to erroneous design while an over-constrained model may unnecessarily restrict the area for problem solving.

5.4.2 Dimension Related Tolerancing

This text does not explain tolerancing as an activity of mechanical design; it just outlines the advanced description and analysis of tolerances in integrated part and assembly models. Because manufacturing processes have some level of errors and more precise manufacturing is much more expensive, dimensions must be defined with a tolerance specification in accordance with the requirements of the operation and manufacturing of the product. However, tolerance cannot be evaluated for an individual dimension but for purposeful chains of dimensions. For this purpose, dimensions are grouped in chains with component and resultant dimensions.

The specification and model description of tolerances need highly skilled engineering: both design and manufacturing aspects must be handled. Optimal tolerances keep production costs down while ensuring the proper function of the assembly. Remember that sometimes a small change of tolerance causes a sudden rise of manufacturing costs. Tolerance is considered as a constraint. Coordinated definitions of geometric, dimensional, and tolerance constraints control and maintain shape related specifications for parts and assemblies.

A tolerance model includes 2D or 3D contours or wires with constrained chains of dimensions and tolerance values. It relates contours by tolerances in parts and assemblies. Relationships are defined between the related chains in the levels of parts and assemblies.

Tolerance specifications are applied on wireframe, surface, and solid part model representations. They are also placed in sheet metal part and assembly models. The following modeling functions assist engineers in the systematic creation and placing of correct tolerances in close connection with an existing design and its changes.

Adding tolerances to single parts and parts of assemblies.

Default tolerances for different types of dimensions and constraints.

Placing proposed tolerances at locations demanded by functions of the product by use of an assembly model. Tolerance type is identified according to the selected geometric elements and features.

Tolerancing forced definition of features when they cannot be extracted from the part model.

Creating tolerances by use of constrained sections or wires in a wireframe model. Sections and wires are extracted from the geometric model representation of the part.

Implicit definition of datum constraint.

Tolerance definitions according to ISO, DIN, ANSI, and other *standards*.

Annotating drawings extracted from the tolerance model. Syntactically correct tolerance description for documentation.

Important *tolerance analyses* are listed as follows.

Verification of tolerance values from the point of view of assembly specifications.

Basic tolerance analysis, e.g., computing the minimum and maximum clearance and interference values.

Analysis for the *possibility to assemble parts* using dimension chains with tolerances.

Checking of *integrity of tolerance models*.

Selection of the *reference dimension* based on fully constrained sections. Calculation of the *probability of holding the tolerance of the reference dimension* while varying the tolerances of all other dimensions.

Calculation of the *acceptance rate* for the tolerance model.

Statistical and worst case stack-up for statistical upper and lower tolerances of the reference dimension and worst case combination of the maximum and minimum dimensions, accordingly.

Sensitivity analysis helps engineers to understand relationships within a tolerance model. It reveals relationships between a reference dimension and other dimensions and shows dimensions having a higher impact on the analyzed reference dimension. Following this, only those dimensions are evaluated that are the main contributors to the reference dimension. The percent contribution of each dimension to the reference dimension can be calculated. This is the basis of improving the tolerance system by loosening tolerances on low contributors, and tightening tolerances on high contributors.

5.5   Animated Shapes

A common misconception about animation is that it means moving objects in a space. The correct definition of animation is that it is a *time-phased change* of one or more *attributes* of an animated object. The basic principle of computer animation is similar to the 2D picture based animation. A similar method was applied in the traditional creation of cartoons where a series of 2D pictures were used as phase information. The sequence of phase pictures in time constituted the cartoon.

Animation of engineering objects is based on geometric models. Selected parameters of an animated object are changed in time through distinct phases are called frames. A series of frames is generated for each animation process. A changed parameter is

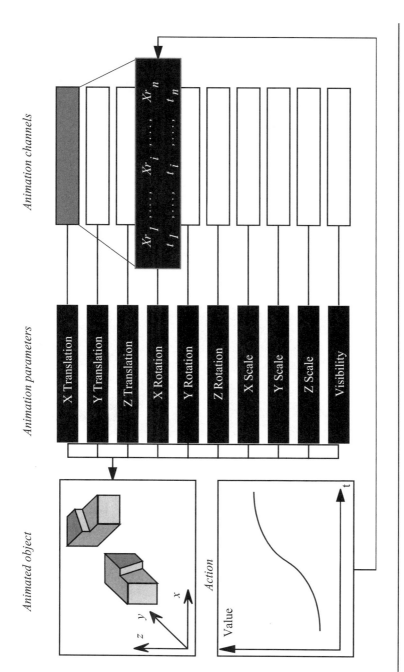

Figure 5-26   Animation of a shape.

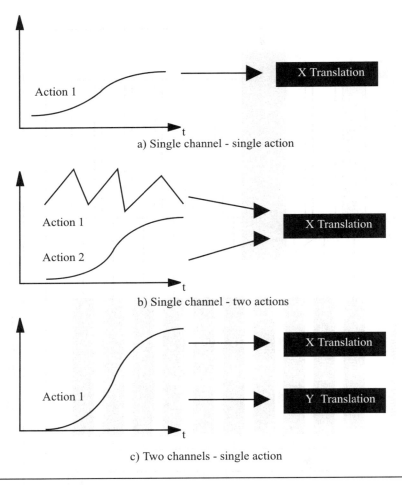

a) Single channel - single action

b) Single channel - two actions

c) Two channels - single action

Figure 5-27 Mapping actions to animation channels.

called an *animation parameter*. Numerous parameters of objects can be animated. The primary application in engineering is the animation of the position of objects in model space. Translation, rotation, and scale may each act as animation parameters in three coordinate directions. For enhanced visual reality, the visibility of objects, position of light sources, and color and intensity of light can be animated. An object that has one or more animation parameters is an *animatable object*. A separate animation channel is defined for each active animation parameter. The animation channel represents how the value of the animated parameter changes over time. The animation process is controlled by discrete values of animation parameters at selected times (Figure 5-26). An animation channel is a set of values for the related animation parameter (Xr_1, etc.) at different times (t_1, etc.).

The change of value of animation parameters over time is controlled by actions (Figure 5-27a). An action contains information for a change in general; it is not created for an animation parameter whereas it is attached to it. As a benefit of this method, an animation parameter change can be controlled by several actions simultaneously (Figure 5-27b) or several animation parameters can be controlled by the same action (Figure 5-27c). A simple definition of action is a 2D curve. A NURBS curve has the capability of excellent local modification necessary for creating actions.

Important applications of animation in engineering are the time-programmed moving of objects along paths and the change of any shapes of objects. The motion path of objects is controlled by motion path action. A single curve as motion path action controls three coordinates. A shape can be animated by creating a series of interpolation shapes between an initial and a final shape. An obvious method in engineering is the change of position of one or more control vertices of a surface. This animation is also cited as metamorphosis. Note that the topology and the number of control vertices in both parameter directions must be unchanged during the animation process.

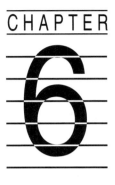

Finite Element and Manufacturing Process Models

6.1 Finite Element Modeling

This chapter presents an overview of *modeling by finite elements* (FE) for analysis of part and assembly models. The theory and application of this method for problem solving in different areas of engineering in static and dynamical as well as linear and nonlinear tasks of deformation, temperature, vibration, frequency, shape, etc., analyses are not discussed in this text.[1] This material concentrates on the less published topic of *FE models and advanced modeling procedures, as they are available for advanced analyses* in Computer Aided Engineering (CAE) systems. CAE emphasizes the analysis based development of products while CAD/CAM

---

[1]Detailed theory and application can be found in: Cook, R. D., Malkus, D. S., Plesha, M. E., Witt, R. J. "Concepts and Applications of Finite Element Analysis," Wiley Text Books, New York, 2001; Chandrupatala, T. R. "Introduction to Finite Element in Engineering," Prentice Hall, Englewood Cliffs, NJ, 1991.

refers to shape modeling and model based manufacturing. Comprehensive CAD/CAM systems offer general-purpose CAE.

Now, the method of finite elements has become a fundamental method of the model based development of products. It reveals the impact that design variables have on design performance. It relies on the calculation of location dependent parameters for *rods*, *shells*, and *volumes* to find areas with critical values for a comprehensive range of analyzed parameters. It is a new and important background for decisions whether to *accept*, *reject*, or *redesign* a part or an assembly during product development. Its first application was in the analysis of aircraft structural elements. The FE method was first published in 1960. By the end of the 1960s *nonlinear problem solving* had been found and by the end of the 1970s the *mathematical basics* were established. These days, important CAD/CAM systems involve FEM and FEA functionality and provide interfaces to FEA systems for more sophisticated analyses in specialized CAE systems.

FE analysis is a *numerical method* for the simulation of behaviors of engineering objects. It is an *approximation*, because the distance between two calculation points of the analyzed *parameter* is finite. As a general problem-solving tool for shapes of arbitrary complexity, it allowed extension of the integration of product modeling to FEM and FEA. The analysis is done on a *finite number of finite elements*. The dimension of the elements is such that the *density of the mesh* depends on the *task* and the *region* of the analyzed shape. The values of the analyzed parameters are calculated using *mathematical equations built into FEA programs or defined by users.*

6.1.1   Associative FE Modeling

Dimension driven, parametric modeling is a relatively new achievement in FEM/FEA systems. Parameters defined for the geometry are completed by parameters defined for the FEM and FEA, such as material properties, force, temperature, and parameters of meshes. Two-way associativity can be established between shape modeling and analysis. Loads, restraints mesh elements, and other entities in FEM and FEA can be associated with part geometry and updated automatically with part design changes.

Changed geometry requires new analysis. In advanced active FEM, analysis based optimization generates change proposals for the geometry or modifies the geometry by associativities. Part modeling regenerates the geometry according to the change demand, and sends the modified part model to a new analysis where new solutions are generated. This iterative process is applied to refine design. *Design criteria* are handled and modified as parameters. Analysis depends on local standards, experience, and knowledge so that special emphasis is placed on *customization*. Engineers applying FEM and FEA systems can define any parameter, entity, or procedure that has an influence on the analysis.

Physical properties such as thickness and cross-section as well as mechanical, thermal, and rheological *material properties* are stored in databases for standard parts, sections, and materials. Properties can be associated with geometry, and then they can be extracted from part models. Most material properties depend on certain parameters as described by mathematical functions. When properties are available for combinations of different parameters, they are recorded in tables.

Geometry is accessed for analysis in part and assembly models. As an alternative, simple shapes can be constructed using geometric tools in FEM systems. Often FEA requires modification of the geometry represented in the part model for design. In other words, the geometry optimal for design differs from the geometry optimal for analysis. Entities representing details of shape and indifferent to analysis may be *deleted* or *suppressed*. On the other hand, points, curves, surfaces, and planes are *added* to the part model as reference entities mainly to control the mesh density or to modify the shape to be meshed. Surfaces that are not included explicitly in the part model may be necessary for mesh generation. Regions of original surfaces, as well as offset, extended, divided, joined, and intersected surfaces are applied for meshing. Surfaces also can be extracted from the solid part model.

Considering the topology only for the geometry necessary for the FE analysis allows generation of a mesh on a collection of surfaces that represents the simplified shape of the part. This method can replace direct modification of part geometry for FE analysis. Wireframe geometry is applied when it carries enough information. Edges, sections, or the surface for the mesh can be composed of multiple curves to simplify the geometry.

Some shells have uniform thickness; others have different shapes on their inner and outer surfaces. For accurate results, the mesh is generated on the *midsurface* between the inner and outer surfaces. Some advanced FEM systems offer automated functions to extract the midsurface and to calculate the shell thickness for calculation. These parameters are handled as physical properties of the shell.

Elements and their relationships in an FE model are shown in Figure 6-1. The part model and the assembly model together with their modifications for FEM and material data constitute the *shape and contact information* for the FE model. The second structural unit of FEM consists of the *mesh* and *information about finite elements*. The environment of the analyzed shape is modeled in the *load model* where loads and boundary conditions are described. Boundary conditions are restraints and supports at connections between parts as they are described in the assembly model. Less advanced systems cannot get contact information from the assembly model. In this case, single part models should be applied and the boundary conditions are defined by the engineer. Material data, finite element definitions, meshes, and load model element definitions are stored in databases as *background information and knowledge* for the creation of FE models. Complete FE models can be stored, retrieved, and applied with the same or modified geometry and load model.

The load model must represent real world conditions. Load modeling is a critical task in FEM because analysis using a formally correct, but inaccurate load model may lead to a result that will be accepted but mistaken. It may suggest development of poor quality even dangerous products. Complicated loads and

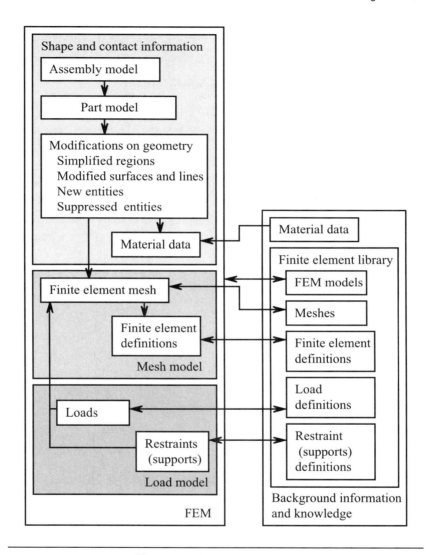

Figure 6-1   Finite element model.

boundary conditions are to be expressed mathematically then described in the model. Loads and supports are defined by the engineer or calculated by computing and analytical procedures. Loads and restraints can be *defined* on and *associated* with part geometry. This method facilitates maintaining *geometry based loads and restraints* during changes of associative design geometry. Loads can be constant or can vary in time. *Multiple load* and restraint sets are defined and stored.

Loads can be placed on points, edges, curves, surfaces, or surfaces fitted to points or sections of a shape. They may be independent of the mesh but applied to nodes and elements. Functional variations of loads are handled by mathematical expressions. *Types of loads* are:

Normal and traction pressures on faces and edges.
Distributed and concentrated forces along the element length and at nodes.
Gravity, translation, and rotation types of acceleration.
Ambient and reference temperatures.
Temperatures at nodes, elements, and geometry.
Heat transfer.
Nodal and distributed heat sources.
Face and edge heat convection, and radiation.

Methods for the description of constraints and restraints are:

Standard restraints: pin (1R), slider (1T), and ball (3R).
Definition of degrees of freedom from 3T and 3R for the so-called virtual part.
Nodal displacement.
Automatic definition of geometry based contact (2T) using the assembly model.

The time-consuming and troublesome manual adding of contacts was replaced by *automatic contact recognition* between parts in advanced systems. Manual editing of the contact surfaces and specifying of the type of contact conditions are available, however.

6.1.2   Mesh Generation

A choice of *finite elements* is available to perform the *allowed types of meshing* and analyses in industrial FEM/FEA modules and systems. Depending on the main characteristics of the meshed geometry, and the requirements of the analysis, one-dimensional (Figure 6-2a), 2D or planar (Figure 6-2b), 3D shell or surface (Figure 6-2c), and solid (Figure 6-2d) elements are applied as building blocks of the mesh. The mesh consists of nodes (V) and edges (E) of elements.

Linear, cubic, quadratic, or even higher degree edges are allowed. If they ensure enough accuracy for the analysis, linear edges are allowed even if the part geometry is curved. Figure 6-3 shows examples for *one-dimensional* (Figure 6-3a), *planar* (Figure 6-3b), *shell* (Figure 6-3c), and *solid* (Figure 6-3d) elements, with

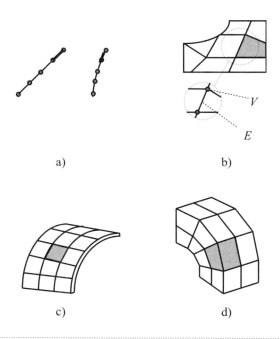

a)   b)

c)   d)

Figure 6-2   Finite element meshes.

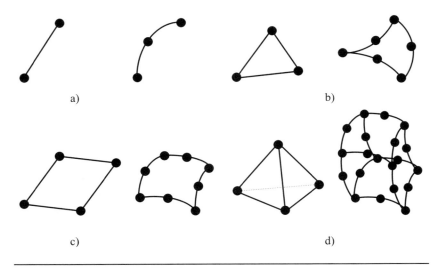

Figure 6-3   Linear and parabolic finite elements.

linear and parabolic edges. As explained for curve modeling, a quadratic edge is defined by three points while a cubic edge must have four points. Polynomial edge representation is applied in parametric meshes. A solid tetrahedron element is called a *P-element* and fits accurately with even fifth order part geometry. Scalar or vector elements can be defined.

The shape of linear, parabolic, or cubic shell elements (Figure 6-4a) can be triangular, quadrilateral, or axisymmetric. Basic solid elements are tetrahedral, cubic wedge, hexahedral or brick, and axisymmetric (Figure 6-4b). Other types are composite, mass, rigid, viscous damping, shear panel, and heat boundary elements.

Erroneous, incomplete, incorrect, or non-optimal FE models, especially meshes, must be corrected. However, the checking of FE models needs sophisticated automatic procedures. Some errors cannot be detected by the usual checking tools in FEM systems.

Verification of FE models includes automatic smoothing of the mesh for minimum element distortion, coincident node and edge checking to eliminate duplications, free-edge and free-face

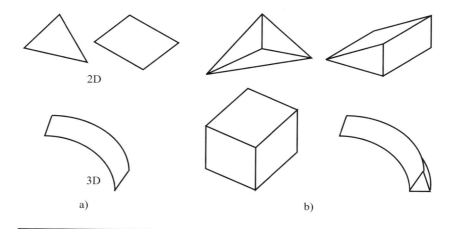

Figure 6-4   Finite element by its shape.

checking to avoid unwanted cracks, and checking for the correct orientation of the normal of the shell. Other checks are for distortion, warping, and stretch of elements. Several automatic procedures are available for optimizing the model, especially the mesh.

By this point, most preprocessor activities have been surveyed and discussed (Figure 6-5). Mesh generation is inevitably at the center of FEM. It serves single or multiple analyses. At *single analysis*, a change of the value of any input variable initiates complete re-mesh and re-solve. In the case of a *what-if analysis*, several input variables are varied in a given range in the space of the input variables. This method organizes analyses. Their number is often very high; various methods are applied to decrease this high number of analyses. The *approximation* type of analysis generates parameters as a function of a range of variables. Application of the *parametric mesh* automatically adapts an initial mesh to modified geometry and load data by modification of nodal coordinates with unchanged element connections and boundary conditions.

A stored or imported mesh or complete FE model can be applied instead of generating a new mesh for each analysis. This possibility must be considered because complicated, time

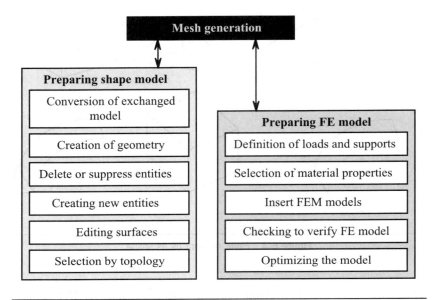

Figure 6-5 Preprocessing.

consuming, and troublesome mesh generation for each new analysis can be avoided. An existing finite element model is modified for a new simulation. The mesh is upgraded for changed geometry or boundary conditions manually, or by parametric and associative features. Design variations are generated and analyzed easily and quickly. Previous experiences can be utilized in existing part and FE models. Several FEM systems can prepare the FE model for multiple applications for structural, electromagnetic modal, etc., analysis.

When appropriately processed geometry is available and the selection of element type and mesh density is ready, *mesh generation* can be done automatically, adaptively, manually, or by mixed application of these methods. Automatic local density definition is based on the curvature and proximity of geometric features. The engineer controls this process by specified values on points, curves or surfaces.

Automatic mesh generation can work on curves and surfaces, as well as solids containing holes and cavities. Automatic minimal element distortion taking allowed element distortion into consideration is an advanced function. The engineer controls the generation process by global and local density specification, and selection of the points, curves, and surfaces to be meshed. *Automatic transition* is generated *between different densities.* Procedures are capable of recognizing reference geometry for mesh definition and creating starting or anchor nodes to control the positioning of nodes.

Adaptive meshing is an automatic modification of mesh density, element order, and element shape according to accuracy and other specifications. The analysis starts with a coarse mesh and then adaptive meshing refines the mesh only in those areas where the specified accuracy needs higher density on the basis of error analysis. Adaptive meshing modifies the mesh by dividing elements, displacing nodes, modification of adaptive P-elements, and even complete re-meshing. The element number can be coordinated on the opposite sides of a shell.

Modification of a mesh can be best automated by the application of the novel *parametric mesh.* The polynomial representation of a mesh can be modified by changed parameters. Typical or task related values of parameters can be stored in databases.

One of the bottom-up approaches is manually controlled meshing by use of *geometry.* The mesh is created on line, wireframe, surface, and solid entities by direct definition of its parameters. Free or mapped mesh generation is applied.

The other bottom-up approach is the *direct creation and editing of the mesh.* Nodes and elements are created in FEM, directly, independent of geometry. Nodes and elements can be created and modified individually. Nodes are created along curves. Other functions are the creation of elements by rotating or tabulating, and transformations of nodes and elements. Elements can be copied, cut, combined, replaced, and moved. Their connections can be modified where necessary. Manual modification of certain models is necessary because it is impossible to develop software with the capability of handling all possible circumstances at mesh generation. On modification of existing meshes, associativites are saved automatically.

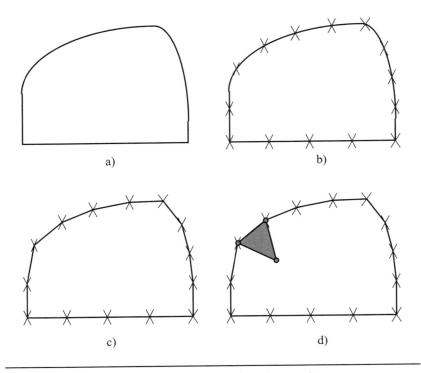

Figure 6-6   Discretizing a shape for a mesh.

The steps of the generation process for a simple mesh explain the basic activities in Figure 6-6. A plane curve as a wireframe (Figure 6-6a) represents the geometry for 2D meshing. The contour is divided according to the density of the mesh (Figure 6-6b). The resultant points will be the places of the nodes on the curve. Because the task allows it, the curved segments of the contour are replaced by straight lines between the points. Segments of the resulting polygon (Figure 6-6c) will be the linear edges of 2D triangular shell elements (Figure 6-6d).

6.1.3   FE Analysis and Post-processing

Structural, thermal, electromagnetic, and flow analyses represent the main application areas of FEA. According to the type of task, analysis is linear or nonlinear. Linear FEA is useful whenever the stress–strain relationship is linear, and displacements and rotations are small. Linear analyses are static for stress and deformation, and dynamical for natural frequency, mode shape, and linear temperatures. When it must be considered that stress, material behavior, contact conditions, or structural stiffness depend on the displacement or temperature, the analysis is nonlinear. Nonlinear analyses are, for example, plastic deformation and strain hardening. Automated solution selection activates the appropriate solver according to the type of problem to be solved.

Amongst the special applications of FE analysis are beam finite elements which serve the analysis of structures where connected beams are applied to carry loads. Structural elements that are long with respect to their other dimensions are represented using the beam finite elements. Very complicated, time-consuming, and troublesome conventional beam design can be replaced by this efficient and relative simple method. Beam elements carry *cross-sectional* information including geometry, moment of inertia, etc., and *connectivity geometry* data. Beam cross-section properties can be defined as an entity in databases. Beam, rod, and pipe elements refer to cross-section tables as well as material and physical properties.

Post-processing supports engineers in understanding analysis results. It provides purposeful processing and graphics displays for FEM and FEA data. Design criteria, loads, restraints, and results are displayed in formatted numerical form, as vector and tensor data, or by color coding of intervals for analyzed parameter data. Results representing displacement, or varying in time, can be animated. Areas can be selected and visualized by selective display where the analyzed parameters have values in a specified range. Results are displayed for load combinations.

Design optimization is an active application of FEA. Design parameters are modified by automatic processing of FEA data to improve the performance of the product. *Design parameters* are optimized according to *design goals* considering *design limits*. Optimization of shapes applies dimensions of simple shapes as

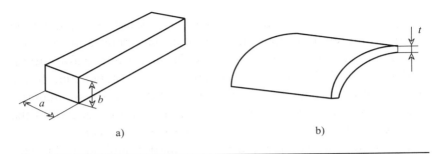

a)   b)

Figure 6-7   Optimized shapes.

design parameters, such as thickness of a shell and cross-section dimensions of a beam (Figure 6-7). Design limits can be the maximum allowable and the minimal required values of design parameters. Design goals are minimum, maximum, or optimal values of performance parameters. For example, two cross-sectional dimensions of a rectangular beam are optimized (Figure 6-7a), where the maximum allowed stress value is the design limit and the design goal is minimum mass of the beam.

6.2   Manufacturing Process Model

This section presents some of the machining process planning and the related analysis as an application of part models in the *integrated, associative modeling of parts, manufacturing processes, and the manufacturing environment.* Principles, processes, equipment, and tools of computer aided machining are detailed only to the extent required for the understanding of modeling tasks.[2] Most shapes are copied from part models into models of tools carrying the shape of the parts. These tools are applied in forming, casting, and other non-machining processes. Milling is the prevailing

---

[2]For related concepts and methods refer to: Chang, T. C., Wysk, R. A., and Wang, H. P. "Computer-Aided Manufacturing," Prentice Hall, Englewood Cliffs, NJ, 1997; Kief, H. B., and Waters, T. F. "Computer Numerical Control," McGraw-Hill, New York, 1992.

primary shape-adding process in their manufacture. *Primary shape adding* is the process that converts the ready-made and analyzed shape model into a physical shape. This is why concepts and methods of problem solving by machining process modeling are discussed through examples from *process milling* in this text.

6.2.1 Model Based Planning of Manufacturing Process

The *procedure of manufacturing process planning* as it is integrated in modeling can be broken down into the activities illustrated in Figure 6-8. Information about the part to be machined is available in the part model as form feature definitions and geometric model representations. A manufacturing process model is developed for a part or a part of one of the tools for its manufacture. Entities in the *part manufacturing process* model represent the process itself, and its elements, including *setup, operation* and *tool cycle*. Setup is a subprocess for a machining in a given clamping position on a machine tool. Operation is a machining within a setup by a given cutting tool. Finally, tool cycle is for machining one or more form features or geometric elements in an uninterrupted sequence of tool paths.

The process includes machining to convert blank or stock pieces to the part according to design parameters and specifications. The machining process and strategy to generate tool paths are selected for each operation. Tooling is configured by its elements, such as cutting tools, tool holders, and other accessories considering tool changing structural elements of machine tools. Manufacturing and economical objectives are fulfilled by process planning for constraints and other design specifications. Manufacturing considerations are process, strategy, number of controlled axes, shape of tool, etc. Economical criteria are minimum allowable material removal rate, minimal cost, or minimal production time.

Tool cycles are created within operations using part geometry and machine tool model information. Tool paths are checked by

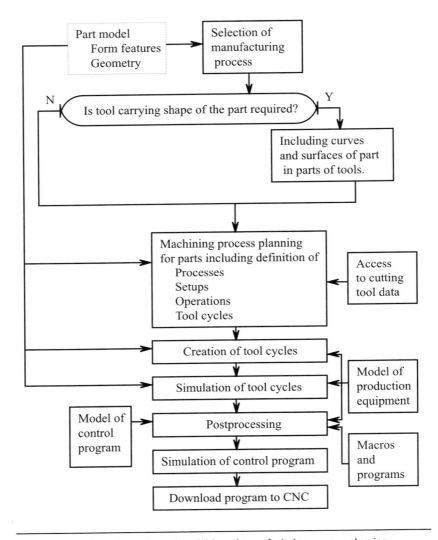

Figure 6-8   Procedure of model based manufacturing process planning.

simulation for feasibility on the selected machine tool, following the tool geometry within the specified tolerances, and collision in the working envelope of the machine tool. Controlled axis information is available in a simplified structural model of the selected machine tool for feasibility checking. Collision checking requires solid models of moving volumes of the part, cutting tool, tool holder, fixture, and machine.

Post-processing produces a control program for the machining of a part under computer control. Macros and approved program details are included in the program; even full programs are retrieved and edited at this stage of process planning. The model of the word address numerical control program includes block, word, and syntax related information. The program is simulated and then downloaded to job floor level into the control unit of the selected machine tool.

Because part and assembly modeling and analysis are available in the integrated modeling system during manufacturing planning, blank pieces, special tools, tool holders, and fixtures can be designed as soon as their design parameters are available.

Part, blank, fixture, manufacturing process, and tool path models can be *associative* in advanced systems. Part and blank geometry changes make associated tool paths ineffective or tool paths are updated in response to part geometry changes. The effects of changes in part and blank design on fixtures and other accessories are handled similarly.

The goal of fundamental simulations is to avoid three main errors (Figure 6-9). Correctness of the selected strategy and the *ability of the machine tool to be positioned to the programmed location* are checked. Controlled axes cannot cope with the tool path if their number and simultaneous control capability are insufficient. The tool path in Figure 6-9 needs two more simultaneously controllable axes. The most frequent cause of *collision* is interference between a tool and a detail of the part during execution of a tool path. *Gouging* is interference between the tool and the machined surface and is checked around the circumference and along the length of the tool. Other goals of simulations are

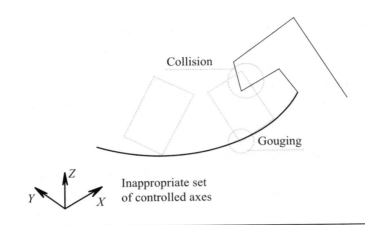

Figure 6-9 Errors in NC programs.

maintaining the specified tolerances, optimizing tool cycles, and minimizing tool wear. Accuracy specifications generally need double precision data.

Real world objects are involved in design and manufacturing engineering as physical shapes to be captured, manufactured parts to be inspected, and rapid prototypes to be visually checked (Figure 6-10). The capture of existing physical shapes during reverse engineering is done according to a measurement plan including points to be grouped in clouds, etc. Inspection of parts proceeds according to measurement inspection plans. Coordinate measurement machines (CMM) are advanced equipment for the measurement of shapes under the control of a programmable control unit. A CMM can be considered as a reverse numerically controlled machine tool where existing part dimension data are produced instead of producing a machined part according to dimension data.

Rapid prototyping through stereolithography (STL) or other processes needs special shape information in the form of layers. The boundary of the shape of the part is processed by special methods, such as approximation by triangles. Control equipment for rapid prototyping receives geometry in standard STL format. The part model is translated into STL. All geometric

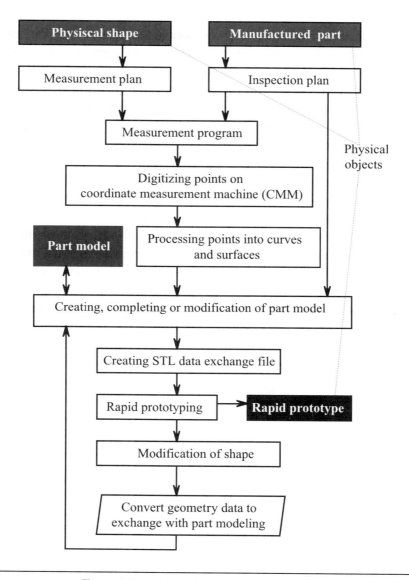

Figure 6-10   Real world objects in shape modeling.

representations including surfaces, solids, and clouds of points can be processed. The part can be divided into separate pieces for making rapid prototypes. Rapid prototyping creates parts by, among other processes, ultraviolet laser solidification of resin or laser cutting from sheets. A reverse shaped rapid prototype of a part with a hard and heat resistant layer on its working surfaces can be applied as a tool for injection molding of plastics. Advanced rapid prototyping control units allow modification of parts and the shape can be developed through several iterations. STL data of the final shape can be translated back into the geometric model for the modification of the original part model.

Setup and operation part manufacturing process model entities are created and captured in the process model mainly manually, with minor but important computer assistance. Some process planning activities, such as selection of machine tools, configuration of setups, selection of cutting tools, and sequencing of operations are assisted by automatic computer procedures in advanced systems. Some of the automatic functions for machining are:

Determining the *operations and machining parameters* needed by a feature.

Tool selection for roughing or finishing of a given form feature, considering dimensions, material, surface quality, and tolerance specifications.

Machining to prevent deflection of slim tools and thin-walled parts.

Reordering the operations to optimize the operation order of machining, and tool changes to minimize the machine time. Several typical methods make sequences by the shortest distance between holes, by tool types, and by increasing tool diameter.

Optimization of the operation sequence by use of precedence rules.

Calculation of in-process shapes for initial operation sequence and reordered operations.

Calculation of depth cut, considering allowances, critical faces, islands, tool related limitations, predicted forces, chip removal, etc.
Feed rate control.

6.2.2   Shape Model Driven Associative Computer Control of Production Machines

The *scenario of the entire computer controlled machining system* is shown by Figure 6-11. The machine tool and machining process are described in a model space with coordinates X_m, Y_m, and Z_m. For programming, a local coordinate system is applied with coordinates X_p, Y_p, and Z_p. The machine tool has five simultaneously controllable axes: X_c, Y_c, Z_c, B_c, C_c. Part coordinates are transformed into controlled coordinates automatically. The

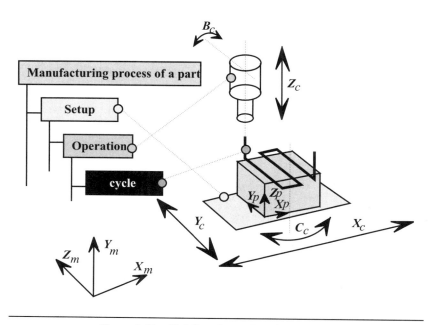

Figure 6-11   Modeling of manufacturing process.

sequence of relative movements of the part and tool is the tool cycle.

Tool paths are calculated from *part model information*. Complete and unambiguous representation of the part is required. This is ensured by previous checks and analyses of the geometry. A geometric entity or a group of entities carry information for the machining task to be solved as a cycle. Recent systems can process form features instead of requiring direct access to geometry. Basic types of geometric entities for the definition of an elementary machining task are as follows:

> *Planar open or closed contours* to be followed by a tool (Figure 6-12a). The controlled point of the tool is on its axis so that the tool path is calculated as an offset of the contour by half the tool diameter. The tool diameter is limited by the smallest inner radius. When the contour involves a curve, its smallest local curvature determines the maximum tool diameter.
>
> Flat surface (Figure 6-12b). When the tool is much smaller than the width of the surface, a strategy is applied to compose the tool cycle.[3]
>
> A *spatial curve* is interpolated to generate the tool path (Figure 6-12c).
>
> The *volume to be removed* is divided into elementary volumes to be removed by moving the tool in a plane. In the case of Figure 6-12d, machining is done on two levels.
>
> *Blank geometry and part geometry* for planning the machining to remove the excess material between them (Figure 6-12e).
>
> *3D surface.* Tool paths are generated along curves in the surface[4] (Figure 6-12f).

Machining multiple features, maybe on multiple parts, in the same cycle by use of the same cutting tool is an efficient

---

[3]For more details see Figure 6-11.
[4]For more details see Figure 6-18.

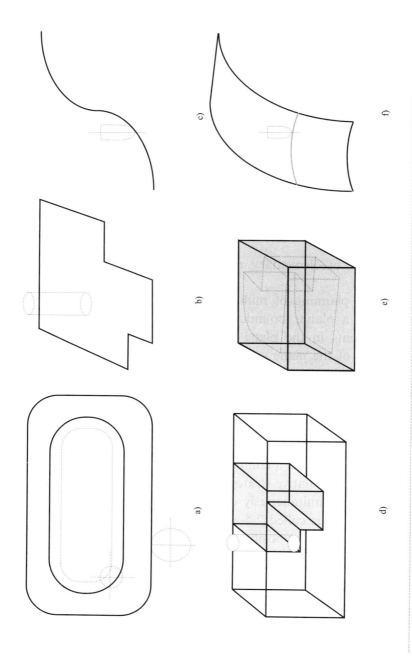

Figure 6-12 Basic entity types for definition of the machining task.

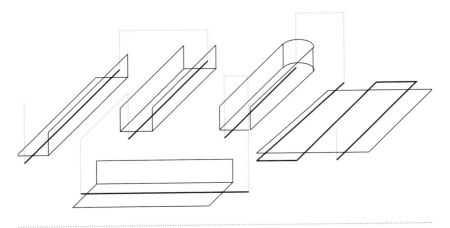

Figure 6-13   Machining several form features in a single cycle.

concentration of elementary machining tasks in a single cycle (Figure 6-13).

For the planning of milling on a plain surface, the simple geometry of a planar boundary contour is sufficient when the tool movements in the plane are not restricted. The geometric environment of the plane surface to be machined produces some limitations against tool movement. Typical limiting effects of environmental geometry are illustrated in Figure 6-14. The tool in the machining of flat surface S_1 is not allowed to move behind walls W_1 and W_2: it also must avoid island I_1. For economic machining, rapid feed is applied over the pocket Po_1. The strategy of the machining is a combination of zigzag St_{1a} and contour milling St_{1-3b}. Alternative strategies are St_2–St_4. Machining can be solved by two simultaneously controlled axes (X and Y) in two dimensions. This is called 2.5-axis control because the third axis (Z) is used but only for positioning above the plane of the machining. Movement of the tool above the clearance plane P_c is guaranteed to be without any risk of collision. Automatic calculation of the working plane on part geometry is a very important feature of cycle modeling procedures.

One of the most problematic and time-consuming activities in process planning is the calculation of intermediate shapes

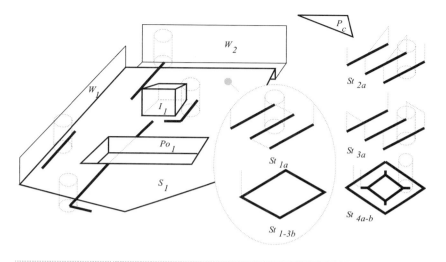

Figure 6-14 Geometry for machining in a plane.

produced at each step of machining. These are the *in-process shapes* and they carry very important geometric information for the subsequent setups and operations. Advanced modeling systems are capable of producing a series of models representing the workpiece shape after each machining operation (Figure 6-15).

When the shape of a blank piece and a final part is very different and a large amount of material must be removed at a high material removal rate, an appropriate volume removal strategy, together with automatic depth-of-cut calculation ensures economical machining. Three main strategies for the removal of material from inner and outer areas of the same part are illustrated in Figures 6-16a–c. The third one (Figure 6-16c) shows one of the passes with the in-process shape after the pass. This strategy is applied for the rough cutting of a part from a box-like blank material in a single cycle in the example of Figure 6-16d.

The machining of press tools of large sheet metal parts such as car parts or injection molding tools of large plastic products needs the removal of a large quantity of material to produce high quality 3D surfaces. Productivity is enhanced by reducing the control of

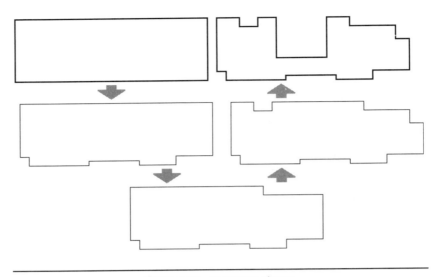

Figure 6-15 In-process shapes.

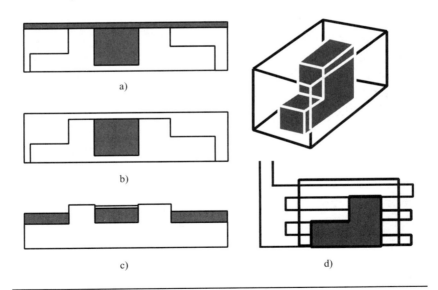

Figure 6-16 Volume removal strategies for prismatic shapes.

machining in a plane. Three different solutions are shown in Figure 6-17. The terracing produces a stepped surface for finishing. This strategy needs only two simultaneously controlled axes in the plane of machining (Figure 6-17a). When a curved roughed surface is needed, a contouring after each 2D pass requires simultaneous control of three linear axes (Figure 6-17b). The third strategy is dividing the material to be removed equally amongst passes in each section along the tool path (Figure 6-17c).

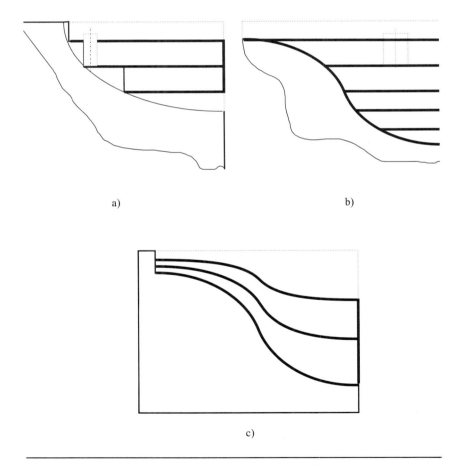

a)

b)

c)

Figure 6-17 Volume removal strategies for 3D surfaces.

A surface is machined through fewer or more tool paths because the tool diameter usually is much smaller than the width of the surface to be machined. Several strategies are applied to generate the proper set of curves for the cycle. Tool paths are isoparametric curves (C_i) or they are generated along the surface by considering a tool that is guided by a series of parallel planes (Pl_i) intersecting the surface (Figure 6-18a).

Curvature analysis is necessary to decide the maximal allowable tool diameter. If the minimum curvature on a surface results in a tool application that cannot fulfill the material removal rate limit, an appropriate change of the shape is to be considered. Another solution might be the application of a larger diameter tool for most of the machining for a higher material removal rate. Then a smaller diameter tool is applied for removal of the material in regions where the curvature is lower than the radius of the economical tool. The procedure analyzes the regions where the tool is not able to reach the surface and generates points at the ends of contacting curve segments. These points are used to create a subsequent surface machining with a smaller cutting tool. As an example, the second pass in Figure 6-18b is done by the tool T_2 between the automatically calculated points P_1 and P_2.

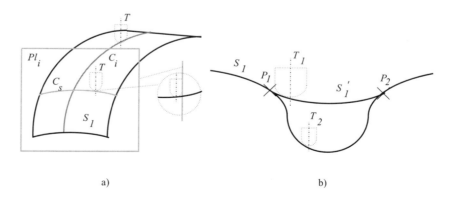

a)   b)

Figure 6-18 Geometry for machining in three dimensions.

Finishing of 3D surfaces can be done with a fixed direction of the tool axis. Machining with a fixed tool axis needs simultaneous control of three linear axes (X, Y, and Z); it is also called 3-axis milling. Its main drawback is the scallop that is a rib shaped protrusion between two tool paths. The height of the scallop on a given surface (H_s in Figure 6-19a) depends on the tool diameter and the distance of the two adjacent tool paths. This distance is the stepover and can be fixed as a percentage of the tool diameter or according to the specified scallop height. If necessary and if both the task and control allow it, the stepover can be variable according to the actual scallop height along the surface.

Control of the tool axis by its rotation around two axes simultaneously with the three linear axes makes the setting of the tool axis direction to the surface normal possible at each point of the surface. In the practice of this 5-axes milling, the axis of the tool is set at a small angle to the surface normal (Figure 6-19b). For machining of drafted plane or curved surfaces, tilting the tool axis around only one axis is enough. This is 4-axes milling where the peripheral of the tool works.

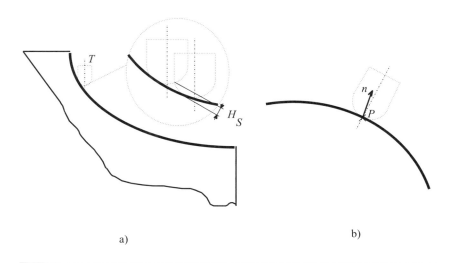

a)   b)

Figure 6-19   Machining of the surface by fixed and controlled angle of the tool axis.

Styling avoids sudden changes of curvatures along well-engineered surfaces. However, certain surfaces contain small and large curvature regions with connection of the specified continuity. The tool radius cannot be larger than the smallest curvature of the surface if the strategy illustrated in Figure 6-18 should be applied. Figure 6-20a shows machining by a tool with the maximum diameter allowed. The controlled point of the tool (P_c) is along its axis. The relation of the controlled point of the tool and the surface normal can be followed along the tool path. Because the path of the controlled point of the tool is the offset of a curve on the surface, its intersection or looping must be avoided (Figure 6-20b). A surface has a specified tolerance range in industrial applications (Figure 6-20c). Consequently, if the tool diameter is larger than the smallest curvature along the curve on the surface but the resulting surface is within the tolerance range, it is accepted. The structure of a tool between its controlled point and the edges can produce geometric relations hard to evaluate. An advanced method compares the theoretical and the machined surface by an analysis. The results are regions where the tool is predicted not to reach the surface, and other regions where the machining is predicted to be below the theoretical surface. Following this, an automatic or manual correction of the tool paths or dividing the machining into two passes with different tools as in the case of Figure 6-18b may be decided.

Let us now consider a few special surface machining strategies. Often a selected region of a surface is to be machined. Figure 6-21a shows tool paths that are generated by interpolation between curves C_1 and C_2 on surface S_1. The number of tool paths along the width of the surface is independent of the distance of the curves. A curve (C_p) can be projected onto the surface (S_p) and then concentric curves as tool paths are generated within the projected boundary of the surface (Figure 6-21c). Figure 6-21b shows radial, along parallel contours and spiral strategies for machining a rotational surface by multi-axis milling.

A *multiple-surface* can be machined in a cycle when the machining can be solved by using a single cutting tool. Entities

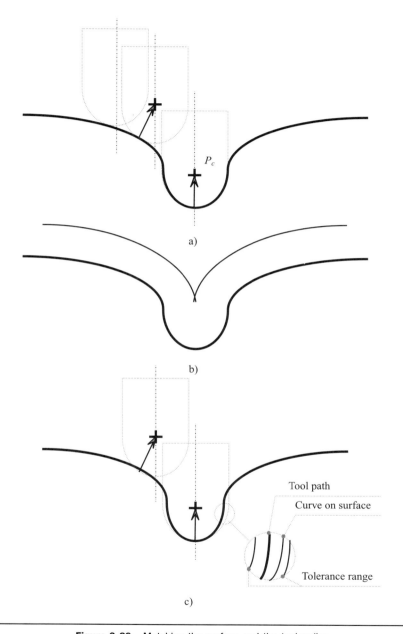

Figure 6-20 Matching the surface and the tool radius.

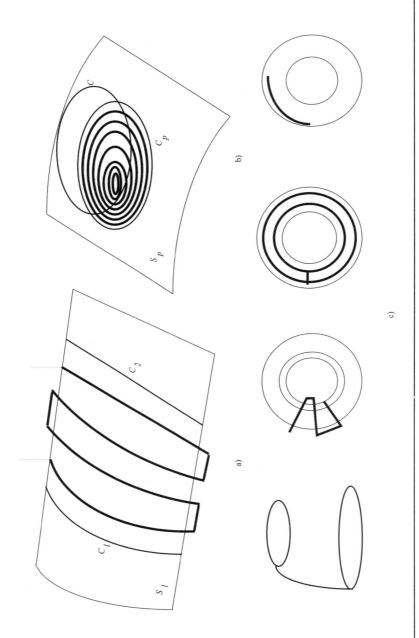

Figure 6-21 Special cycles for machining surfaces.

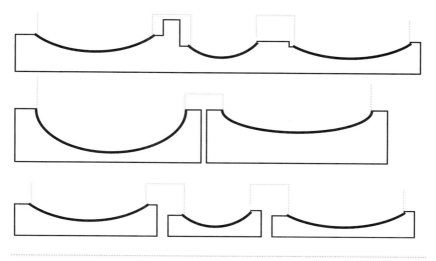

Figure 6-22   Connecting tool paths.

can be grouped by manual selection or an automatic grouping procedure. Machining of a multiple-surface and of *surfaces on a multiple-model* is illustrated in Figure 6-22. The path of the tool from its end at one surface to the start on the next surface is called a connection or auxiliary motion. The tool is retracted to avoid collision in the area between the surfaces. Because it is generally not necessary to bring the tool to the clearance plane, it can be more economical to calculate the height of retraction from the model description of the obstacles between the surfaces.

Simulations for checking of the feasibility of a cycle process input information about the controllable axes, and the relationship between axial movements and volumes moving in the working space of the machine tool. For this purpose, a simplified model is appropriate. Figure 6-23 introduces such a model by the example of a typical milling machine. In this example, the directions of the model space coordinates and the controlled axes are the same. In the case of a flexible manufacturing system with several machine tools in the model space, the directions of the controlled axes are described by vectors. The simplified model describes the direction and extent of slide movements, the structure of the machine tool,

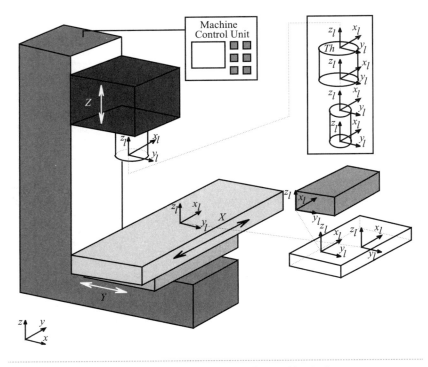

Figure 6-23 Simplified model of a machine tool.

and the position of all structural and manufacturing process dependent shape units. The structure of the machine tool describes the chain of slides between the table for the workpiece and the spindle for the tool. In the example of Figure 6-23, a frame connects slides for controlled movements Y and Z. The slide for movement X is carried by the slide for movement Y, so that movement Y moves slide X. Machining task dependent elements are the tool, tool holder, fixture, and workpiece. They are interconnected and connected to the machine tool by matching local coordinate systems.

Post-processing receives tool movement and part geometry information in the form of an Automatically Programmed Tool (APT) or Cutter Location File (CLFILE). Because post-processing can modify part movements, bi-directional translation is necessary between tool cycle generation and post-processing.

6.2.3 Simulations and Post-processing

The post-processor receives tool cycle information in a form independent of the applied control unit–machine tool pair. It produces a control program in the *symbolic word address programming language* of the selected control unit and fits process parameters such as the speed, feed rate, etc. to the selected machine tool. Repeated details, such as connecting motions and programs for frequent form features can be included in the control program as parametric macros. They need only the setting of the actual values of parameters as dimensions, etc. Larger units, even full programs, are also included. The application of proven and verified program segments and programs reduces programming and testing time and risk of errors.

A post-processor is created or customized in the application environment for a single or several similar machine control units by use of information about the machine tool, control unit, and local standards. General-purpose macros are written at the development of the post-processor programs. User-defined macros have the same importance where quick modification and reuse of similar control programs is necessary. A post-processor creates control program blocks automatically, using machine control unit specific program language information such as words and syntax.

Machining can be simulated before, during, and after post-processing. The tool path can be traced, modified, and verified with the support of visualized geometry. Shaded solids ensure excellent possibility to check areas where the tool cannot reach the surface or gouges into it. It is obvious that the programmer must be able to modify the program as necessary at this stage of process planning. Simple graphical methods are available to add, transform, delete, trim, and move tool paths and their points.

A basic choice of an advanced set of analyses and simulations can be organized in groups for checking the *surface model, machine independent program, control program,* and *economical consequences* (Figure 6-24). *Interactive graphic* simulation together

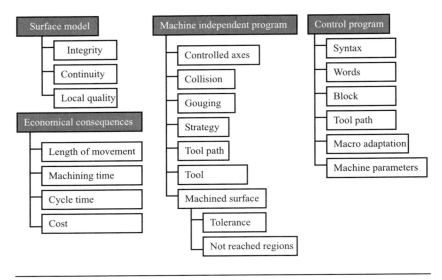

Figure 6-24   Advanced analyses and simulations.

with advanced *visualization* helps the engineer in finding errors and defining the best sequence of machining and tools.

The *surface model* is analyzed before its application for creating tool cycles for integrity of surface, continuity as specified for manufacturing, and local quality.

Controlled axes, strategy, collision, and gouging are discussed above. Other checking of *machine independent program*s includes preliminary what-if analysis to evaluate the effects of different tool options, following cycles step-by-step, and prediction of the position of the machined surface in relation to the modeled surface.

Exact graphical visualization of tool path movement with display of the geometric model and in-process shapes can be applied on entire programs, cycles, or even individual tool movements. When the part is mounted onto a machine tool model in the model space, simulation can be set for a fixed or moving workpiece. Some simple cycles are defined in the control program by only their parameters to calculate tool paths by procedures built into the machine control unit. Simulation performs calculation tool paths of these cycles.

The *control program* is checked for syntax, words, program block composition, tool path definition, macro adaptation, and application or actual machine tool parameters. Specifications for the program language, machine control unit, and machine tool are applied.

Some analyses ensure economical process plans and control programs. The *economical consequences of a process plan and control program* can be assessed by cost estimation based on machining time, total cycle time, length of tool paths, number of cutting tools, etc.

Creating Curve and Surface Models in CAD/CAM Systems

Chapters 7–9 give an organized and detailed description and explanation of techniques from the present advanced *integrated product description based modeling* technology. The text discusses important model construction processes by their operation modes, means for human control, form, and content of input and relations to other construction processes.

Humans and computer procedures interact to develop more complete or less complete computer descriptions of products depending on the extent of the engineering activities to be assisted by computer model based engineering sessions. Computer descriptions of engineering objects are stored as data sets and consist of integrated or stand-alone data subsets. Integrated subsets are interconnected and model data flow is inherently ensured amongst them. Stand-alone subsets in most cases describe the same information in different formats. Moreover, the capabilities for the description of the same information sets may be very different. Interconnections of stand-alone subsets often require data transfer with conversion of content and format or even manually assisted data transfer.

Interconnection of modeling activities using *stand-alone data sets* is illustrated by general (Figure 7-1a) and example (Figure 7-1b) schemes. The data subset is generated by a *stand-alone modeling procedure subset*. It cannot be processed directly by other modeling procedure subsets. Figure 7-1a shows the case of application of neutral data exchange files by double conversion. An alternative

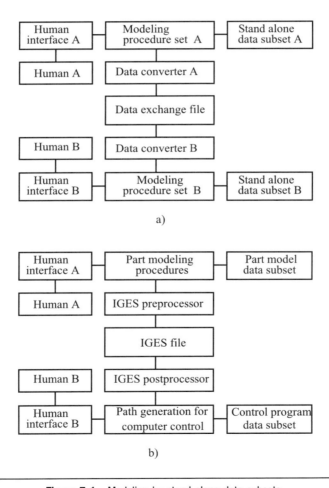

Figure 7-1   Modeling by stand-alone data subsets.

solution would be the direct conversion to a data set suitable for modeling procedure subset B. Modeling procedure subset B produces model data subset B. Relationships are not described between data subsets A and B. Figure 7-1b illustrates the application of stand-alone modeling procedure and data subsets by the example of part modeling and part manufacturing planning. Modeling sessions are interconnected by an IGES[1] data exchange file.

Application of stand-alone model data subsets restricts possibilities for human–human, human–procedure and procedure–procedure communications, automated information processing, and evaluation of model descriptions. It is time consuming and there is a constant risk of losing important information. The handling of the effects of change of a data subset on other data subsets is complicated. Consequently, consistency of a comprehensive product model is not easy to evaluate. Modeling by stand-alone data subsets cannot be considered as advanced.

Integrated product data descriptions relate interconnected data sets in a single, unified, and well-structured database (Figure 7-2). Modeling procedures put and get information in the same format. They can understand and process each other's data without any conversion. A single and unified human interface serves all engineering activities.

Integrated computer descriptions involve the interrelation of different engineering objects and their attributes. This is one of the most valuable features of integrated modeling. Each modeling procedure accesses any related information directly and produces new or modified information accessible by any other of the procedures. The modeling procedures understand information without any conversions or human assistance and define new relationship information for existing and new information. Humans define

---

[1]IGES (Initial Graphics Exchange Specification) is the most widely applied standard file format for the exchange of geometric model data between dissimilar modeling systems. All CAD/CAM systems include programs for translation of their own model data format into IGES format and IGES format into their own model data format.

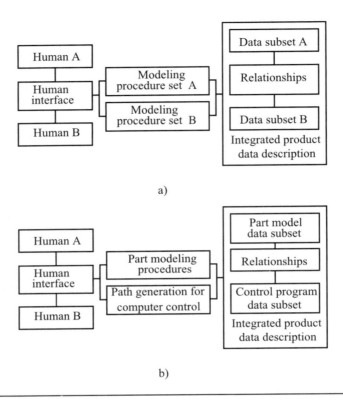

Figure 7-2 Modeling by integrated data subsets.

new model information and control the work of the modeling procedures during two-way communication sessions.

Modeling procedures are created for well-determined modeling functions. An industrial CAD/CAM system involves a more extensive or less extensive set of functions according to its purpose and field of application. CAD/CAM systems compete in offering procedures and representations for all modeling functions necessary for the description of typical products.

7.1 Aspects of Model Creation

Several terms such as concurrent, simultaneous, group worked, process oriented, feature driven, object oriented, project based, behavior based, parametric, and contextual are applied as attributes to the words *engineering, design,* and *modeling.* These attributes are *aspects* with same importance in the modeling of engineering objects. Special emphasis on one or more aspects often gives a deformed picture of industrially applied modeling. Engineers involved in product related engineering activities consider several aspects in a consistent and harmonic innovation process. Aspects of model creation are as follows:

> *Context of entity.* Engineering objects are modeled as independent objects or in context with other objects. Similarly, model objects are defined individually or in some *context.* Context defines a hierarchic relationship between modeled objects or model entities. By use of this relationship, the related object or entity is applied as input information at the creation of a new or modified object definition or model entity.
>
> *Project.* Engineering objects are often modeled in *project based* organizations. A project is established for a well-defined group of engineering tasks. It may cover a single product, a group of similar products, or a family of products, structural units, or parts. Other projects are organized around orders from customers. Some structural units and parts are applied in different projects and special projects involve purchased structural units and parts. Models that are applied in one project but created in another project are referred to but not included in the applying project.
>
> *Engineering object to be described.* The subjects of modeling are engineering objects: this is why aspects of the engineering objects are so important. Successful modeling describes engineering objects as they are intended by the engineer and informs the engineer about the real behavior of the modeled objects in the virtual world. Finally, successful modeling gives a description of products suitable for the programming of computer-controlled production of parts, as they are conceptualized and detailed by engineers.

Objectives and limitations specified by engineering tasks. The objectives of an engineering activity are considered by humans and engineering procedures. They are as follows:

A function as it is specified by its parameters.

A shape having design engineering, styling, manufacturing engineering, and mathematics related characteristics.

Maximum utilization of a material.

Minimum weight of a part, a structural unit, or a product.

Resistance against a specified effect from the outside world.

Minimal cost and taking the competitive price of the product into consideration.

Objectives are associated with specifications of *limitations* for critical object parameters such as stress, temperature, cost, etc., to avoid definitely unwanted regions of their values.

Behaviors of modeled objects. Product related engineering objects have intended behaviors when they are working in their physical world. One of the purposes of modeling is to describe all the product information necessary to evaluate the behavior of modeled physical engineering objects in the virtual world of the computer by advanced simulation such as FEM/FEA or similar methods. Instead of the "make then check" style of conventional modeling, advanced modeling evaluates behaviors of engineering objects during the creation of their models and *produces behavioral checked model objects.*

Shape to be described. The shape of a part of an engineering object is defined by using functional, aesthetic, and manu-facturing ideas and specifications. Design of a mechanical system is *shape centered.* All other information can be attached to shape information.

Model representation. Model representation depends on the set of information to be described and related. It is also determined by the capabilities of the modeling systems for creation, understanding, and application of models.

Variants of modeled objects. Products are designed in more or fewer variants, depending on their application area and customer demands. Modeling methods are developed to support the design of variants. Variants are best represented in a single model to avoid redundancies and an unnecessarily complex model structure. The possibility for a quick change between variants is called flexibility. Flexible engineering is allowed by variant based modeling, while flexible production is allowed by flexible manufacturing. Flexible technology is based on the integration of engineering, manufacturing, production, marketing, and customer services in a computer system.

Design (human) intent. Information about an engineering object originates in the human brain in the form of ideas and concepts. The concluded information is communicated with modeling procedures as the design intent. This intent suffers less or more distortion before it is established as a computer representation. The computer system processes information and communicates it with other humans and equipment control procedures. The most important question to be answered at evaluation of engineering activities is how the product agrees with the original human intent for it.

Construction of the model. Construction of a model is the implementation of an engineering process in an industrial modeling system. It depends on the information available about and necessary for the modeled engineering objects. Modeling as a powerful tool for a successful engineering process, is evaluated on the basis of its ability to process and describe proper information about engineering objects. Sets of construction tools are available in modeling systems for creating and manipulating elementary and combined shapes and other models. Interactive procedures enable engineers to make quick, associative, and intuitive modeling.

Process of engineering activities. The innovation process of a product starts from the first idea and ends at the last improvements of the production of its final version. Product related engineering activities move beyond this point for customer and end-of-life related affairs. Engineering activities for design, analysis, manufacturing, production, marketing, and customer services can be sequential, concurrent, or networked. Integrated modeling connects engineering subprocesses into an integrated engineering process, where an elementary process is active when minimal information necessary for any step within it is available as a result of other engineering subprocesses. The appropriate organization of processes is *concurrent engineering* where engineering subprocesses overlap. The output of an engineering subprocess and its application as input of other engineering subprocesses are governed by modeled relationships amongst engineering object descriptions. This advanced modeling has been established during the past decade and is often cited as *process oriented modeling*. This means that an engineering subprocess is defined and realized in the context of other engineering subprocesses on the basis of contextual relations of the modeled engineering objects. Consequently, process orientation of the modeling is realized indirectly, through relations of modeled objects.

Group work of engineers. Groups and teams of engineers are organized and managed for related groups of modeling tasks. Each member accesses projects and subprojects and can do engineering activities according to the *roles* accepted for that member and the actual *states* of the engineering objects. A role refers to an engineering subprocess managed by a group of authorized engineers. For example, an engineer has access to cylindrical parts in a subproject for a gearbox for detailed design of parts having the status *under detailed design*. When the status of a part is changed to *under acceptance process*, the engineer cannot access it any more.

After an initial understanding of an engineering task, several questions should be answered by engineers as follows:

What initial steps are optimal for the computer description of the first concepts and engineering objects?
What initial information is to be considered?
How is initial information best acquired from human and computer sources?
What additional steps of modeling are necessary for the description of forthcoming engineering objects?
What additional information is needed for systematic construction of the model to the final result?
How is additional information best acquired from human and computer sources?

The above questions can be answered only with the knowledge of hundreds of model creation methods applied to solve everyday and special modeling tasks.

7.2 Curve Models

Lines and curves are considered as initial elements in shape model construction. Higher level entities, such as wire frames, surfaces, solid primitives, and form features, are defined in the context of simple or compound lines and curves.

7.2.1 Construction of Line and Curve Models

Most line and curve definitions rely on *point definitions*. Points are defined in the context of a coordinate system (P_d), a curve (P_c), or a surface (P_s). A point is a geometric entity as defined in the STEP (Figure 7-3).

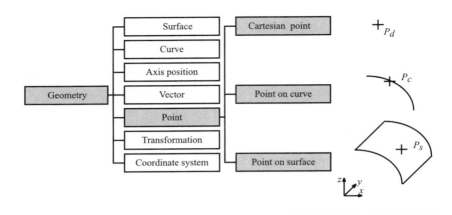

Figure 7-3 Definition of points.

A point is defined for the creation of a curve or it is taken from an existing geometric model by picking a point on a graphical screen. When a curve, e.g., a circle, is defined by one or more of its parameters, such as center point and diameter, attribute values are used instead of points.

The definition of a free form curve entity applies control points (Figure 7-4a), interpolation points (Figure 7-4b), or a hand sketch (Figure 7-4c) as input information. Unevenly spaced points are fitted accurately by correct modeling procedures. A hand sketch is processed into a harmonic curve of the similar shape.

Analytic lines and curves are defined by type, type related parameters, and characteristic points. In Figure 7-5a, an *arc* is defined by its ends, P_1 and P_2, and its radius. Three segments of a compound arc in Figure 7-5b are specified by different values of their radii (r_1–r_3). Segment connections P_1 and P_2 are specified as fixed points.

When lines are defined in the contexts of points, lines, and circles, the situations outlined in Figure 7-6 are the most frequent. A line may be at a distance from a line or at an angle with a line. Sometimes it goes through two points. A hand sketch approximates a shape intended by an engineer: it is processed into the specified or best-fit analytic curve. A line can be tangent to a

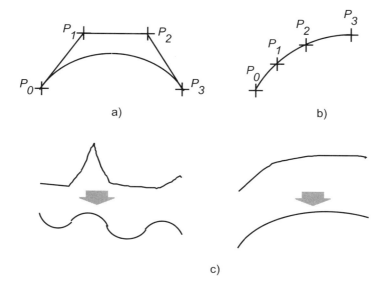

Figure 7-4 Definition of a curve.

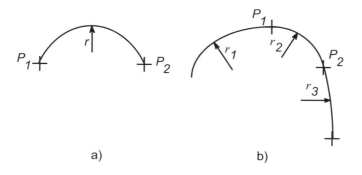

Figure 7-5 Definition of single and compound arcs.

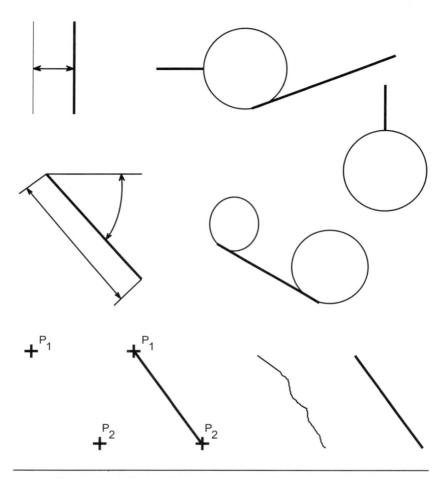

Figure 7-6 Definition of lines in the context of other lines and circles.

circle, tangent from a circle to another circle, or perpendicular to a circle.

An *arc* can be defined by using any three of the following parameters: two end points, another point on the arc and a center point, or the radius. Several methods of arc and circle definitions in the context of existing arcs and circles and a hand sketch are illustrated in Figure 7-7. An arc may have ends common with an existing curve (Figure 7-7a). Circles may be concentric with

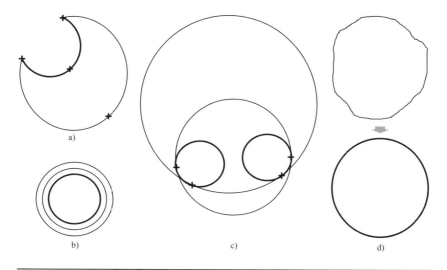

Figure 7-7   Definition of circles in the context of other circles.

existing circles (Figure 7-7b) or may touch two existing circles (Figure 7-7c). A circle is generated following a hand sketch if the entity is specified as a circle or the hand sketch has a best fit to a circle (Figure 7-7d).

An ellipse is created using the length of the major or minor axis, the axis length ratio, the center point, and the sweep angle as parameters.

When a curve is defined *in the context of existing entities*, it is created

> using points taken from existing entities (Figure 7-8),
> as a characteristic curve on a surface (Figure 7-9), or
> as a transformed copy or an offset of a curve (Figure 7-12).

Points for the definition of a curve may be selected as individual points as well as points on a curve, a contour, or a surface (Figure 7-8a). Curve C_1 in Figure 7-8b was created through four points of surface S_1. Curve C_1 is guaranteed to lie in the surface S_1 only at the interpolation points. Between two points, the curve certainly must be pulled onto the surface S_1.

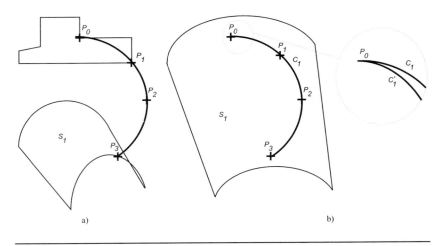

Figure 7-8   Definition of curve in the context of points on lines and surfaces.

There are various relationship definitions applied at the generation of a new curve in the context of an existing curve. As an example, a curve can be created with linearly varying distance to the other curve.

Advanced model construction uses the *sketch in place* method for creating a curve on a surface (Figure 7-9). In the world of 3D curves and surfaces, planar curves are still very important. They should be constrained into a plane (Figure 7-9a). A curve on a surface can be created on-line during the definition of approximated or interpolated points on a surface (Figure 7-9b). A hand sketch on a surface serves both functional openings and styled forms (Figure 7-9c). Open or closed curves may be projected onto the surface from a point or in a direction defined by a vector (Figure 7-9d). All points of the resulting curves are guaranteed to lie on the parent surface.

During and after their creation, *characteristic curves* can be created on surfaces according to their type. Figure 7-10 shows examples of curves extracted from surfaces as an isoparametric curve (Figure 7-10a), an interpolation curve on a lofted surface (Figure 7-10b), and a section curve on a swept surface

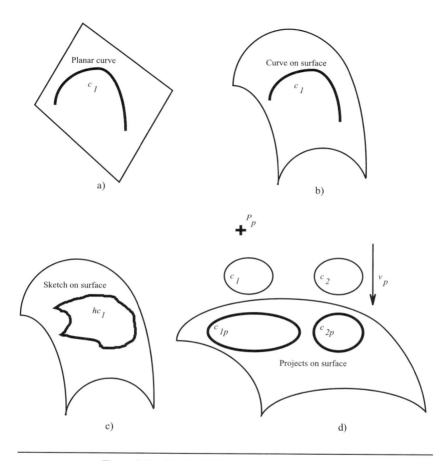

Figure 7-9   Direct definition of a curve on a surface.

(Figure 7-10c). Curves can be extracted from an intersection of two surfaces or an edge of a part (Figure 7-11).

Offsetting and transformation produce controlled copies of curves. Curve C_o on Figure 7-12a is an offset of the curve C. Thin walls can be created between closed contours (Cc_1, Cc_2) and their offsets ((Cc_{1o}, Cc_{2o}), Figure 7-12b, c). Objects are oriented and placed in the desired location by transformations. Curve C_t was created by translation and curve C_{tr} by combined translation and rotation of the curve C. Objects also may be scaled by appropriate transformations.

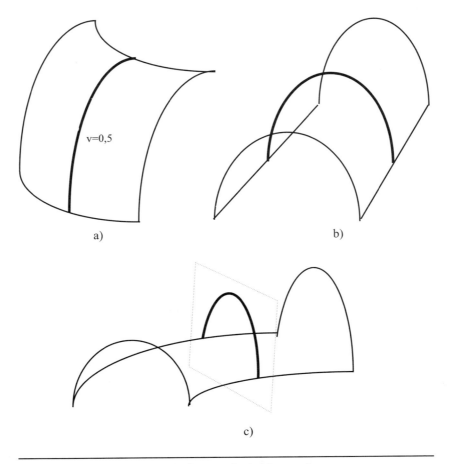

a)

b)

c)

Figure 7-10 Curves extracted from surfaces.

Creation of a curve in one or more contexts constrains the entity. Constraints are to be maintained after any modification. Consequently, they control the curve by an interpolation point, a control point, direction, continuity to a curve or surface, and intersections with curves and surfaces.

The *continuity constraint* can be positional (G0), tangential (G1), curvature (G2), a consistent rate of change of curvature (G3), and the rate of change of the rate of change of the curvature (G4). A tangent at a point can be specified by a direction

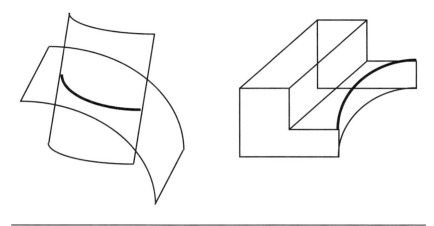

Figure 7-11   Curves extracted from a surface intersection and an edge of a part.

or associatively with another line. Figure 7-13 shows typical examples of continuity constraints. Curve C_i is defined as the connecting curve between curves C_1 and C_2 with second order (curvature, G2) continuity at the connections (Figure 7-13a). Curve C_3 is extended by a line with first order continuity at its end P_3 (Figure 7-13b). A tangent is specified at the point P_4 on curve C_4. The shape of curve C_4 changes according to this new constraint (Figure 7-13c). The shape of curve C_5 is controlled by four planes. The tangent of the curve is the same as the normal of the planes at the selected points on C_5 (Figure 7-13d).

Processing of a hand sketch for open and closed compound lines requires consideration of both entity and associativity definitions. In the example of Figure 7-14, automatic processing of entities E_1 and E_2 recognized straight lines L_1 and L_2 and right angle associativity between them. Ideas are quickly sketched by using a hand sketch functionality. An advantage of hand sketching is construction without interruption for searches on menus and other activities for specification of entities.

Old style curve modeling procedures asked for input of control or interpolation points and then generated the curve. Advanced procedures create curves in *real time* with point definitions.

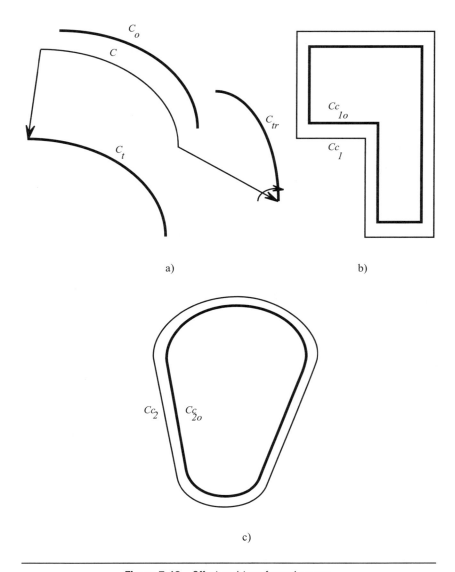

a)

b)

c)

Figure 7-12   Offset and transformed curves.

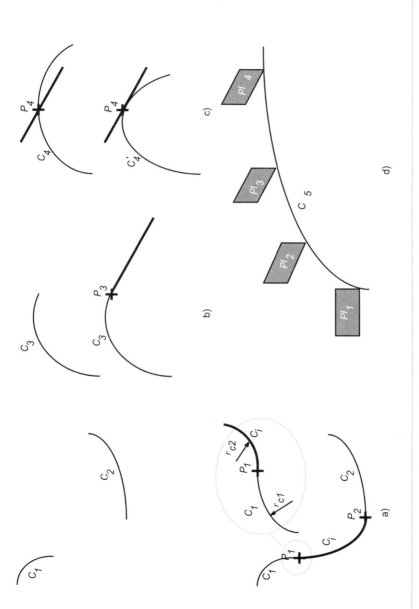

Figure 7-13   Control of curves by contextual continuity specifications.

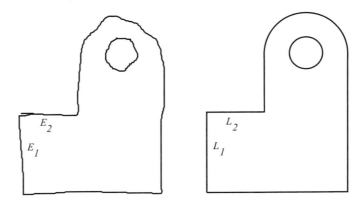

Figure 7-14   Processing of hand sketched closed contours.

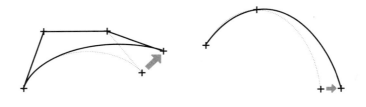

Figure 7-15   Real time generation of a curve.

Moving the temporary end of the approximation or interpolation drags the curve, allowing flexible shape definition (Figure 7-15). The shape of the curve varies with the actual position of the gripped point.

Points can be approximated and interpolated by an *open* or *closed* curve. Interpolation of points P_1 and P_2 by open (Figure 7-16a) and closed (Figure 7-16b) curves produces different shapes of the curve because the constraint of closeness requires smaller curvatures. In Figure 7-16c, an open contour is controlled by constraints at its ends in the form of tangents equal to the same points on the closed curve. It can be seen that the shape of the open curve is forced to that of the closed curve.

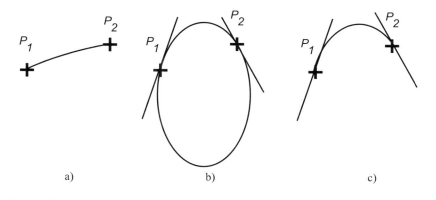

Figure 7-16 Creating open and closed curves.

Open and closed line and curve primitives are available in modeling systems. Open primitives are the line, polyline, arc, hyperbola, parabola, and free form curve. Closed primitives are the rectangle, circle, ellipse, and free form curve. Chains of open primitives play a very important role as open and closed compound lines in the construction of shape models. Chamfers and fillets can be defined at sharp connection points of primitives. The suitability of a compound line as an input entity for surface creation is restricted by the demands of the selected type of surface. A compound line containing break points, for example, cannot be applied at the creation of certain surfaces in certain modeling systems.

A special compound line is the *perforate*. It is applied at tabulation of a solid prism with multiple through holes. In Figure 7-17, closed primitive Cp_o is perforated by closed primitives Cp_1, Cp_2, and Cp_3 for tabulation of solid primitive Ps_1 along vector v_1.

7.2.2 Construction Rules and Navigators

The conventional approach to dimensioning associates dimensions with geometric elements. The presently prevailing *dimension driven* approach defines geometry by dimensions. Geometry is associated

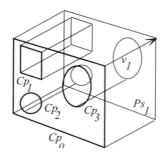

Figure 7-17 A perforate and its application.

with dimensions and other constraints so that any modification to dimensions updates the geometry. Dimension driven modeling of single and compound lines and curves serves as a basis for modeling part families. Other common industry terms for different applications of dimension driven design are *variational geometry* and *parametric design*.

Geometry driven dimensioning and *dimension driven geometry* are compared in Figure 7-18. The compound line in Figure 7-18a was composed using geometric primitives with their own dimensions. The change of line L_1 opened the closed chain because the connecting point of the line entities was not constrained. Line L_2 has to be enlarged accordingly, in a separate operation. The compound line in Figure 7-18b was defined as dimension and constraint driven. Dimensions and end coincidences are specified as constraints. The connecting starts and ends remained coincident, and the shape changed. An additional right angle constraint at the right top corner would produce a conflict situation that could be resolved by deleting the length of L_2 as a constraint. As another example of shape change by a change of dimension, Figure 7-18c shows a change of direction of a contour feature by a change of L_1.

Construction of simple and compound lines and curves is a starting session of shape modeling by wires, surfaces, solid primitives, and form features. Engineers communicate the design in the form of *dimensional, geometric, and logical construction or design rules* for the definition of line primitives and their relationships.

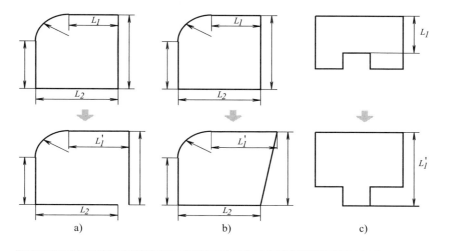

Figure 7-18 Geometry driven dimensioning and dimension driven geometry.

Design and construction attributes of the word *rule* refer to *two aspects*: the design of an engineering object is produced during the construction of a model representing it. Construction rules control the construction of models and carry design information in the form of constraints. They provide a fundamental method for information-rich and effective computer aided design. Construction rules can be communicated by *direct command* or *navigation*. Navigation offers actual design rules for the engineer to select from, considering recent and actual modeling steps and the cursor position. Construction rule information is acquired automatically or by human command.

Construction rules define the shape, dimension, position, and relationships of entities. Existing entities are used for the purpose of specification of one or more characteristics of new entities. Construction rules bring new constraints into the model. Modification of the model must not break valid construction rules.

Dimensional construction rules define the length, radius, distance, or angle. In Figure 7-19, point P_1 is gripped by a pointing device and the engineer may apply four different design rules as offered by a navigator procedure to configure the line primitive

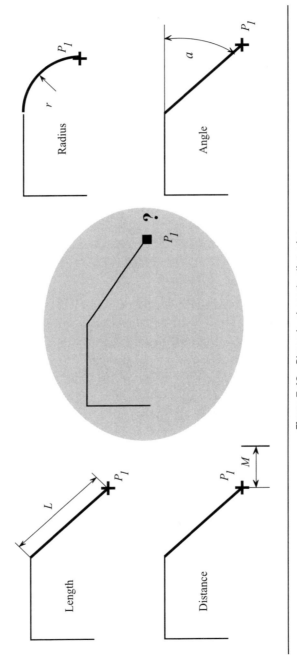

Figure 7-19   Dimensional construction rules.

being defined. The values of dimensions *L*, *M*, *r*, and α can be set and constrained. It can be seen that the dimensions determine the shape.

Geometric construction rules are fixed position, horizontal, vertical, end point, and mid point. In Figure 7-20a choice of four rules is offered by the navigation. One of them should be selected, in contrast to dimensional construction rules. Following this, the straight-line segment can be extended using the rule that its end point will be the mid point of the extended entity.

Logical construction rules are tangent, parallel, perpendicular, concentric, continuous, coincident, symmetric, intersected, point on element, collinear, connect, and equal (Figure 7-21).

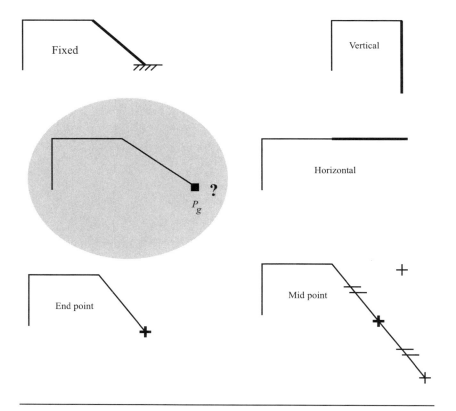

Figure 7-20 Geometric construction rules.

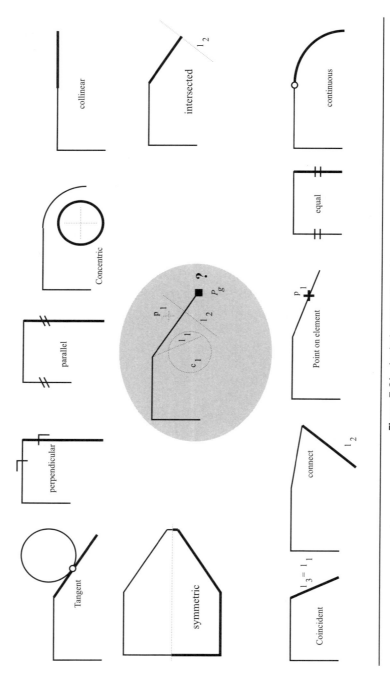

Figure 7-21 Logical construction rules.

Modeling functions in the conventional construction of contours using line and curve primitives are picked from multilevel menus. Advanced modeling applies *navigation* to recognize and offer *possible next steps of construction and construction rules.* Dynamic navigation offers actual construction rules depending on the *actual cursor position.* Only solutions relevant to the actual entity and its environment are offered. Acceptance of a construction rule by clicking at the correct cursor position captures the design intent and places constraints in the model. It makes construction of contours quick and effective. Construction by navigation is assisted by graphical features such as feedback of construction rules by graphical symbols or change of the color or light intensity of lines representing the actual entity.

The area of possible cursor positions is divided into *intent zones* in the view port. A construction rule is attached to each intent zone. In Figure 7-22 five intent zones are communicated with the engineer, who selects intentionally or intuitively. User-defined intent zones can be applied in certain modeling systems. The actually modified element is often called a rubberband because it can be shaped by a very flexible construction method.

It is not necessary to bring the cursor to the correct position; intent is recognized within the predefined *snap zone* (Figure 7-23). Some modeling systems work with the orientation of the actual entity. In this case, the intent zone is defined as the tolerance of the angular position of the entity.

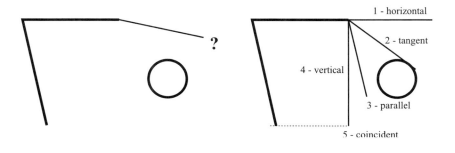

Figure 7-22   Construction rules as they are recognized by navigation.

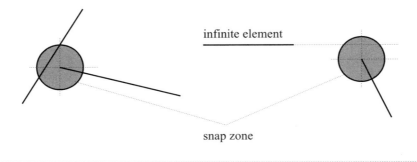

infinite element

snap zone

Figure 7-23 Snap zone.

7.2.3 Modification of Curves

Curves are modified to *change characteristics, modify shape, trim or extend, smooth, and match with the environment.* Construction rules or construction rule related parameters are changed. Line, curve, and compound line related dimensions are modified as for the construction rule related parameters. Construction rules can be deleted, inserted, or suppressed. References are defined to objects. Curves, curve segments, or individual points can be *unlinked from references. References can be changed or deleted.*

Basic curve modifications are summarized in Figure 7-24. The change of position of an approximation or an interpolation point on a curve or the modification of a tangent at a point of the curve changes the shape (Figure 7-24a). Tangent constraints can be modified dynamically or by their numerical value. The change of tangent or position of a point on a curve can be done with or without the change of value of curve parameter u at that point. Individual points or curve segments can be added or deleted. Control and interpolation *points* can be added to the beginning or the end of a curve and can be placed at an arbitrary point. In old style curve modeling, the extension of a curve was done by redefinition of input control vertices or drag points. In advanced curve modeling, the mouse pointer is positioned where the additional point is to be added (Figure 7-24b). The *curve is*

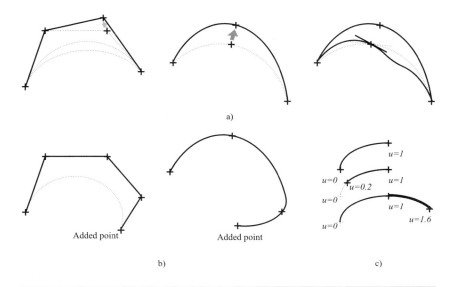

Figure 7-24   Modification of a curve.

extended dynamically, with or without constraining. It is created in real time. The curve follows the newly defined point when it is not a predefined or existing curve. Interpolation points can be moved along the curve. *Knots* are added, deleted, and moved on B-spline curves.

Control points are moved dynamically by a pointing device, with navigation or as a result of changed numerical information. Multiple curves can be modified simultaneously.

The range of curve parameter u can be changed by modifying the start and end values (Figure 7-24c). Parameter values of B-spline curves at approximation or interpolation points are modified to change the curve characteristics. Change of class and continuity conditions of the curve is a means of improvement for a well-engineered shape. The direction of the curve can be reversed when its application demands it; for example, for the construction of tools carrying the shape of a part.

Trimming is widely applied to control the extent and dimensions of curves. The model description saves the original curve information. The original curve can be revealed and used for construction purposes; for example, to intersect with a curve or surface.

Transformations in the model space are explained in Figure 7-25. Shapes are *repositioned, copied, and re-dimensioned* by translation, rotation, mirroring, and scaling transformations and their combinations.

Extensive sets of other construction tools are available for the modification of compound lines: some of them are given in Figures 7-26 to 7-28. Closed contours are *decomposed* into subcontours or line primitives (Figure 7-26). Details of contours are applied in the construction of mating parts, tools, and cutting tool cycles. Lines are *divided and limited* by single and compound line entities (Figure 7-27). Elements of compound lines and polylines can be *deleted, suppressed, and inserted* (Figure 7-28).

Direct modification of curves changes their size, position, or orientation by *dragging* one of the *handles* by a pointing device. End points, mid points, and other user-defined points can serve as handles. A curve can be changed by moving locations on it. In Figure 7-29a, moving the mid point of an arc modifies its radius. Dragging can be applied to complete entities (Figure 7-29b). The simultaneous moving of several dragging points is applied for stretching a single object (Figure 7-29c, d) or multiple objects (Figure 7-29e).

The construction rule is one of the most important modeling tools for saving the design intent. In Figure 7-30, one of the two lines in the right angle is rotated around its connection point. When the construction rule is saved, the other line is forced to rotate with the first line. On the other hand, an intentional new construction rule can be used for the modification of existing entities. In the example of Figure 7-31, modifications of closed compound lines are forced by *equal* and *parallel* construction rules. Figure 7-32 shows three examples where *tangent* and *parallel* construction rules as constraints save the original relationships between model entities after modification.

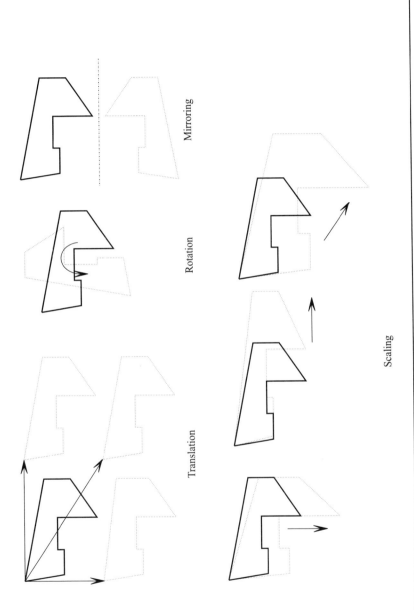

Mirroring

Rotation

Translation

Scaling

Figure 7-25   Transformations.

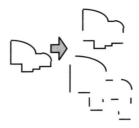

Figure 7-26 Decomposition of a closed contour.

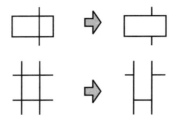

Figure 7-27 Dividing and limiting lines.

Figure 7-28 Deletion and insertion of line elements in a polyline.

7.2.4 Joining and Blending Curves

Joining and blending of two existing curves are applied extensively to make contours that are more complex. Figure 7-33 organizes different situations. When a connection point exists, the result may be a single curve or two curves according to the task. If component curves are applied individually, they must be kept. When the distance between the curves to be connected is small, automatic snapping can be applied. One or both of the curves may be altered to achieve the specified continuity at the connection point. If one of

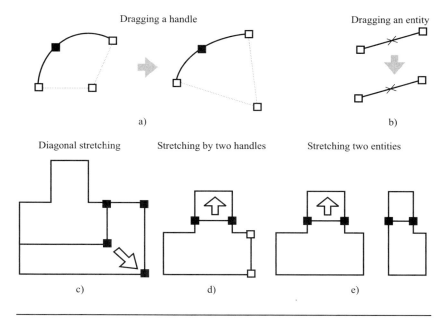

Figure 7-29   Dragging and stretching entities by handles.

Figure 7-30   Modification without and with a construction rule.

the curves is a line, arc, or section of one of the conic curves, its shape cannot be altered, so the second curve must be modified to establish the specified continuity. If the constraint *rational* is removed, then the curve can be altered as a free form curve. Sometimes an additional constraint is that the common point must remain in its original position.

When the curves to be joined have no common point, a third curve is defined to complete the curves. The result can be one, two, or three curves according to the application. The connecting curve

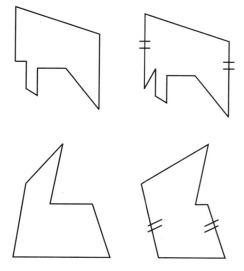

Figure 7-31 Modification by construction rules.

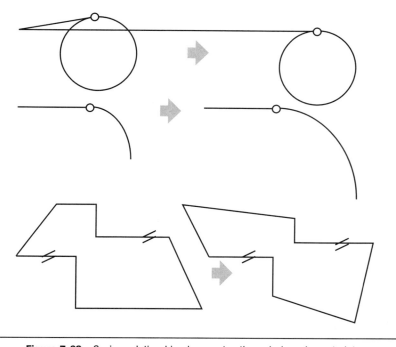

Figure 7-32 Saving relationships by construction rule based constraints.

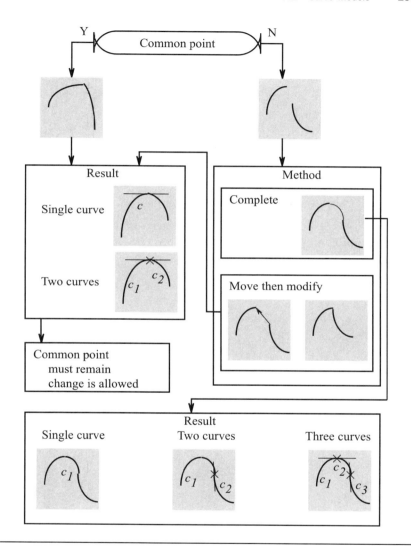

Figure 7-33 Joining curves.

is controlled at its creation to establish the specified continuities at both of the connections. As an alternative, one of the curves can be moved to the other one to establish the common point. After matching their end points by transforming one of them along a vector, the curves are handled as curves having a common point.

When two curves are adjusted to enforce their intersection, and one of them cannot be modified, it is the *target curve*. The other curve will be adjusted to establish a specified intersection. A degree of the curve that is appropriate for the specified continuity is supposed. The degree of the modified curve must be cubic or higher, because this allows for modification of the shape. A lower degree is allowed for the target curve than for the modified curve. Figure 7-34 shows the four basic options for the removal of a

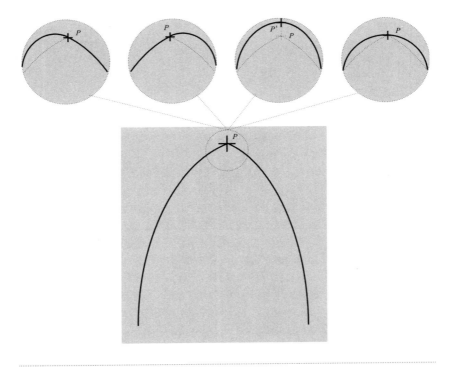

Figure 7-34 Options for the removal of a break at the connection of curves.

break at the connection of two curves. This operation is called *blending* and can be done with the unchanged or changed position of the connection point P as well as with modification of one or both of the connected curves.

When two curves are to be connected by a newly generated third curve, the continuity specifications can be fulfilled by a suitable shape of this third curve. Three-dimensional curves C_1 and C_2 are completed by a third three-dimensional curve C_3 in Figure 7-35b. In Figure 7-35c, connecting curve C_3 is specified as an arc. It is generated with tangent continuity at its connection with C_2; the shape of C_1 must be modified for the continuity (Figure 7-35d). In the case of curvature continuity specification at the connection of C_2 and C_3, curve C_2 also must be modified.

Closed line primitives (Figure 7-36a) can be combined by union (Figure 7-36b), difference (Figure 7-36c), and intersection (Figure 7-36d) operations similarly to constructive solid geometry.

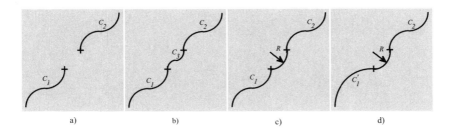

Figure 7-35   Connection of three-dimensional curves.

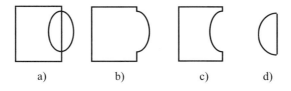

Figure 7-36   Combination of line primitives.

7.2.5   B-Spline Curves

The parameter values along a B-spline curve are assigned to control vertices or interpolated points by a knot spacing function during the creation and modification of the curve. The parameter range is *equally spaced* or parameter values at these points depend on the *position of the point* along the curve. The first method results in a uniform B-spline curve, the second method in a non-uniform B-spline curve. *Uniform parameterization* is easier to handle, while *non-uniform* curves have better *curvature distribution*. The degree of the curve is specified usually between 1 and 8, with the possibility of higher degrees in advanced modeling systems.

When the system in which a curve model is created and any other systems with which that model is exchanged support a *multiplicity factor*, this can be set at a control vertex to a factor up to a limit defined by the degree of the curve. The shape of a segment of a rational B-spline curve can be controlled by increasing or decreasing the weight factor at the actual control vertex. When all control vertices have a *weight factor* of 1, the rational curve behaves as a non-rational one.

7.3   Construction of Surface Models

Conventional modeling applied *separate surface and solid modeling*. Now the modeling of the boundary representation of solids has been integrated with surface modeling. This text follows the presently prevailing *fully integrated approach* where surface models are constructed in special purpose *shape modeling environments* and then are placed in the boundary representation of a solid model for parts of products and tools for their manufacture. *Surface* representations are *associative* to *solid part geometry* and updated with changes to the solid representation of the part.

Model based design of *high quality surfaces* is still a challenge across a wide range of industries in the production of aerodynamic aircraft skins, aesthetic automobile surfaces, hydrodynamic ship hulls, household appliances, computers, electronic entertainment products, furniture, watches, sports equipment, and toys. Modeling procedures are required to reach the specified quality of design by easy and quick model creating processes. Advanced control methods for the *shaping and reshaping* of surfaces cope with the challenging and ever changing objectives of styling and engineering. Imported or scanned geometry should be merged into on-site constructed surface and solid models. The comprehensive surface modeling functionality of CAD/CAM systems follows the ever increasing requirements for the quality of surfaces in leading product design.

Despite the aesthetic oriented nature of surface design, the accuracy of surface description is an important aspect to be considered. A *tolerance* is defined by the design engineer according to the *application* and *purpose* of a surface, and forthcoming *model construction operations* such as its intersection with curves and surfaces. Advanced modeling procedures generate optimal surfaces in accordance with the specified tolerance.

The modeling of complex surfaces is an *iterative* process during which the engineer *creates* a surface model, *analyzes* its characteristics, and then *modifies* it to improve the design and manufacturability. Design engineers, stylists, and manufacturing engineers collaborate on a trade-off of the aesthetic, functionality, and manufacturability aspects.

More and more products are covered by *complex sculptured surfaces* and *multi-surface shapes*. One of the primary objectives in the development of modeling systems is the ability to handle any desired shape regardless of its complexity. The traditional way is by creating single surfaces then *blending* them into complex ones. An advanced solution is the application of a *curve network* as complex input information to create a *set of surfaces* in an associative way. Complex surfaces are often called *skins*.

Surface modeling produces boundary representations of surfaces with or without a *history of model construction*. The development and modification of complex curves and surfaces needs information about how existing curve and surface entities were created and combined. At the same time, some modeling problems can be solved better without a history description. Modeling systems that offer history recording also allow modeling without history.

The construction history of a loft surface and its offset can be followed in Figure 7-37. The entities used in the construction of a surface are called *constructors*. Two section curves and two limiting curves are the constructors of loft surface 1. The offset of loft surface 1 was generated using the offset value. When any of the constructors is changed, the associative surfaces change accordingly. The history of surface model construction provides the link between surface model and the elements that were used for its creation as the constructors. Modification of a surface model is

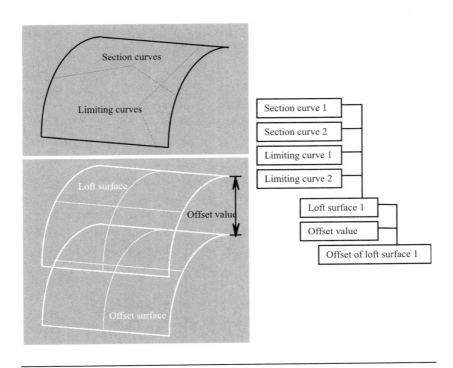

Figure 7-37   History for a loft surface and its offset.

simplified by the possibility to go back in its history to the point where an entity or its attribution changes.

The transformation of objects in the history is not allowed because this activity would contradict the main principle of history based modeling, that changed parameters of entities generate changes of other entities. When moving of the history of an object is needed, the construction history will be lost.

Closed curves represent the outer and inner boundary edges of surfaces. Certain modeling systems allow for trimming by several individual curves enclosing a region. Boundary edges can be used for the generation of new surface entities or for the trimming of existing ones.

7.3.1 Creating Surfaces by Construction Laws

In this section, construction methods are detailed as they are applied in practice. Most surfaces are based on one of the construction laws.[2] Although surfaces in advanced modeling are parametric representations, flat surfaces as mapped to topological faces are often represented without parameterization.

A *revolved surface* is created by revolving a planar curve about an axis according to the start and sweep angles. The radius of the revolved surface can be set by changing the position of the axis. Change of one of the end points of the axis modifies the direction of the axis. A s*ingle curve* or a *set of curves* can be revolved and a *group of revolved surfaces* of the same axis, start angle, and sweep angle or a *set of separate revolved surfaces* may be produced in a single operation. The degree of the surfaces can be equal to or less than the degree of the constructor curves.

Tabulating was originally done along a vector. In advanced modeling systems, extrusion can be done along a curve similarly to

---

[2]The related principles of shape generation and the characteristics of surfaces are detailed in section 3.2.3.3.

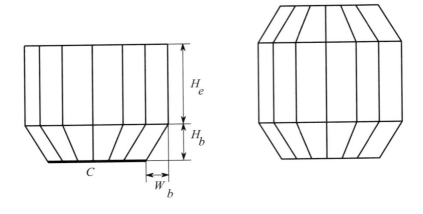

Figure 7-38   Bevel surface.

the sweep surface. The difference is that the spine curve is not applied. The tabulated surface maintains the orientation of the curve, while the orientation of a swept surface varies along the path according to its tangent. A *bevel surface* is created as a combination of tabulation and extrusion, with a single or double sided bevel (Figure 7-38). Starting from the curve C, the bevel is extruded with depth H_b and width W_b. Then the end curve of the bevel is tabulated with depth H_e in the direction of the axis. The step of the bevel extrusion can be repeated on the other side of the tabulated surface starting from the tabulated curve. The ends of the bevel surface can be capped to gain a closed surface for the boundary of a solid. The cap may be a flat surface. As an alternative, a 3D cap surface can be created ensuring tangency continuity at the connection with the bevel.

A *lofted surface*[3] passes through a set of predefined open or closed, planar or non-planar, and evenly or unevenly spaced section curves as construction curves. Its shape is controlled by *surface tangency* at arbitrary points of the construction curves or by *rails* acting as border curves and spines. Rails can be free

---

[3]For the basics of lofting see Figure 3-40.

curves, curves on surfaces, or curves on surface edges. The surface is automatically interpolated between sections, with specified order of the curves in the surface. Curves can be added and removed during and after generation of the surface. When the last curve joins to the first curve, the lofted surface is closed.

There are methods and tricks to the parameterization of lofted surfaces according to its application. Parameterization can be controlled in the u and v directions separately. There are several methods for *knot spacing*. In the direction of surface generation, it can be *uniform* or *non-uniform* on the basis of the average distances between the construction curves or the actual length of the curves. A surface mesh can be defined between any pair of construction curves by specification of the number of spans. Knots can be matched to any of the construction curves. If each curve has this number of knots with the same parameter values, the lofted surface will have the same number of spans along its curves as the construction curves. If the parameterization of construction curves is not matched, the number of spans will be higher than in any of the construction curves.

Versatile application and flexible control of shape have made *swept surfaces* very popular. One or more generation curves are swept along a path curve using one of the sweep modes, according to the selected type of control of generation curves along the path. A minimum of one path and one generation curve are needed and a separate spine curve can be defined.

The relationship between *generation* and *path* curves is defined by the sweep *pivot point*. The pivot acts as the center at rotation and scale of the generation curve. The pivot point can be selected on the generation curve, on the path curve, or off these curves. Figure 7-39 shows possible places of the pivot point. The natural position of the pivot point P_p is on the generation curve C_g, nearest to the path curve C_p (Figure 7-39a). The pivot point can be moved on the generator curve to its arbitrary point (Figure 7-39b) or to its end (Figure 7-39c). Finally, the pivot point can be placed outside of the generation curve (Figure 7-39d). The pivot point can also be moved on the path curve.

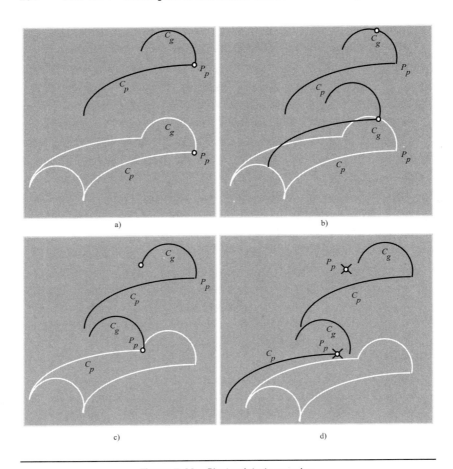

Figure 7-39 Pivot point at sweeping.

The section to be swept or the generation curve can be of any planar shape such as a straight line, arc, conic, free form curve, or series of joined curves. The sweep can be done by only one segment of the generation curve. More than one generation curve can be swept along two rails acting as paths. Two generation curves are applied to define the start and end curves as boundaries of the surface. During generation of the surface, sweeping of the two generation surfaces moves them towards each other. The surfaces are blended at the mid point of the sweep. When the interior area of the surface is to be defined, more than two generation curves can be applied. Section curves can be created along the surface at sample points on the spine. Generation and cross-section curves may be open and non-planar. Each sample point defines a section plane.

Variable section sweep surfaces are constructed by controlling the orientation of cross-sections by scaling and tilting the generation curve in its plane. A generation curve can be scaled in two coordinate directions in its plane. Scaling can reduce the generator curve to a point by the end of the path. An angle specifies linearly increasing tilting of the generation curve along its path.

Sweep can be controlled by multi-rail. Tangency to adjacent surfaces also can be controlled. Multiple rails can be followed by only the position or position and surface edge tangency. A variable radius blend can be created between the sweep and connecting surfaces.

Extrusion, sweep, and other path curve based methods can produce twisting of sections around the path where the direction of the path curve changes suddenly. Including new control vertices or new interpolation points makes the curve more gradual to avoid this condition.

7.3.2 Creating Boundary Surfaces

Surfaces having *three, four, or more sides as their boundaries* can be constructed using separate curves (Figure 7-40a). The order of the selection of curves is not important in the case of boundaries with even numbers of sides, but it is important for boundaries with three and other odd numbers of sides. A boundary with an even number of sides is recommended to be defined in the opposing pair order of same direction curves. The same direction of opposite curves in this case prevents the generated surface from twisting. The *directions* of parameters u and v are often defined by the order of the boundary curves. For example, the first boundary

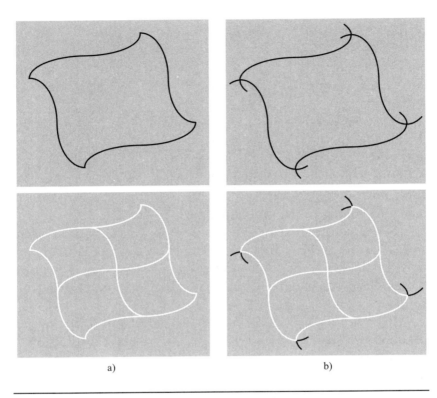

a)   b)

Figure 7-40 Boundary surfaces.

curve represents the direction of parameter u, the second one the direction of parameter v.

Construction curves for boundaries can be extracted from existing surfaces around the boundary surface. In this case, the generation of the surface can maintain continuity with adjacent surfaces automatically. Tangent (G1) or curvature (G2) continuity at the connection of the resultant surface with the adjacent surfaces may be specified for each boundary curve. The *boundary curve can extend beyond its section* used as the boundary of a surface. Construction curves are sections of boundary curves (Figure 7-40b). They can be extracted and then re-parameterized, reducing curve data and improving parameterization.

Curves for a boundary must be intersecting. There is a *tolerance range of intersection* specified for each modeling system within which the specified continuity of the surface is guaranteed. If the *intersection gap* is greater than the tolerance range, the surface perhaps is created but the continuity at its boundaries may not be maintained. The intersection error is recommended to be corrected by moving curves into positions so as to make intersection within the tolerance range.

The creation of a surface for boundary curves is an *interpolation of boundaries* providing the specified tangent and curvature continuity conditions along the surface boundary. The influence of pairs of boundary curves, number of spans, and the rate of blend across the surface between opposite pairs of boundary curves are considered. Spans are inserted until the required continuity is achieved or the specified maximum number of spans is reached. A single surface is created or the resultant surface is a combination of multiple surfaces. The surface is interpolated (Figure 7-40) or created by trimming an initial surface (Figure 7-41). In the case of the method of trimming, surface S_1 is created by using trimmer curve Tc_1. Tc_1 is a curve on the surface S_1. Following this, surface S_1 is trimmed by Tc_1. The degree of the curve can be defined in the u and v parameter directions. Using a higher degree makes it possible to achieve the required continuity with fewer spans.

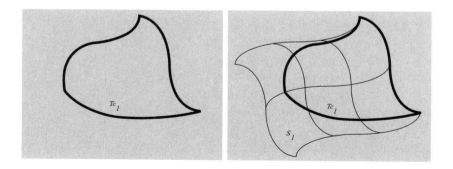

Figure 7-41   Creating a boundary surface by trimming.

The shape can be controlled by a distinguishing point on the surface called its *center*. This point can be moved along the surface, controlling the shape of the surface.

Complex geometry can be constructed using several surfaces in the form of a *composite surface*. A composite surface may be inappropriate for some applications. Component surfaces may not line up at their boundaries and they may be of different degrees. In these cases, its component surfaces should be combined into a single *combine surface*. Sometimes certain component surfaces cannot be included into a combine surface. The combine surface is regenerated so that its degree can be selected. The surface creating procedure generates an outer boundary; this boundary typically consists of four curves. Continuity with adjacent surfaces can be specified. Following this, a single surface is generated within the boundary. The original component surfaces are sampled and the interior of the new surface is modified to fit the new surface into the component surfaces.

7.3.3   Creating Fillet Surfaces

Fillet surface S_f is constructed between two existing surfaces, S_1 and S_2, and can be defined as circular (Figure 7-42) or free form. Tolerances may be specified for contact of the fillet and the surface. Two surfaces having no intersection and the specified radius r are inputs for surface generation. Surfaces have to be close enough that the fillet can be created. Failed surface creation can be corrected by moving the surfaces closer or by using a larger radius.

Free form fillet surfaces are created between two curves on two different surfaces (Figure 7-43). A boundary curve, an isoparametric curve, or a curve on a surface can be applied for this purpose. It is practical to select an isoparametric curve. If an appropriate isoparametric curve cannot be found on the surface, the precision of the curve can be increased as necessary or an isoparametric curve can be generated. In Figure 7-43, isoparametric

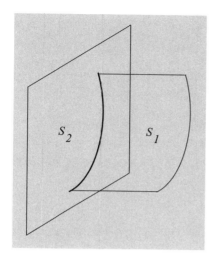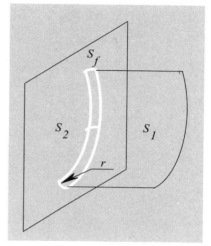

Figure 7-42   Circular fillet surface.

curve Ci_1 on surface S_1 and boundary curve Cb_2 on surface S_2 are applied to connect the surfaces by free form fillet surface S_f.

The shape of the fillet surface can be controlled by several means. The fillet depth is controlled starting from a chamfer, the shape of the fillet can be skewed towards one of its sides. When filleting is defined by an isoparametric curve or curve on a surface, the surface is trimmed back (Figure 7-43b) or not (Figure 7-43c).

Most fillets are placed by *blending along existing edges* of surfaces (Figure 7-44). This operation is sometimes called rounding. The rolling ball principle[4] can be applied. The fillet is required to be tangent with the original surfaces. Fillets are created along isoparametric boundary curves or trimming edges. Fixed or variable fillets are defined by the radius at appropriate points along the edges. The radius also can be defined as varied by a function along the edge. If two edges meet at a vertex (Figure 7-44b), two fillets are generated, then intersected, and finally trimmed back to their intersection boundaries. If three edges meet at a vertex, each with

---

[4]See Figure 4-6.

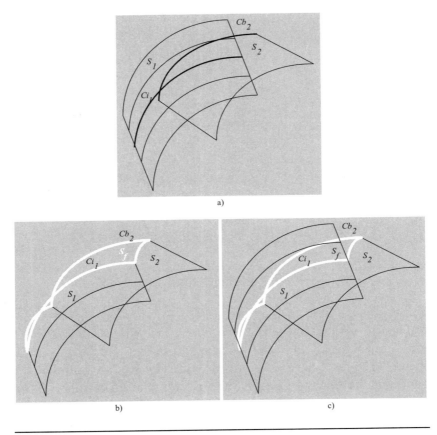

Figure 7-43 Free form fillet surface.

the same radius (Figure 7-44c), a portion of a sphere is created. In the case of different radii, or more than three vertices, smooth filleting is provided by blending the surface with three or four sides. Each fillet edge must intersect adjacent surfaces: if this is not the case, one or more fillets should be extended.

7.3.4 Creating Surfaces Starting from a Curve in a Direction

Surfaces up to this point in the text have been created by intersecting, filleting, or connecting existing curves and surfaces. Sometimes

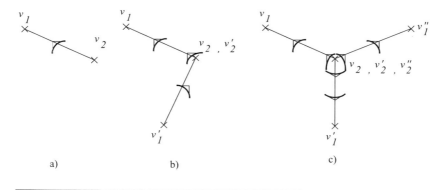

Figure 7-44 Filleting on edges.

surfaces should be created starting from a single curve or curve set in a direction, without any other curve or surface defining its other end.

Single or multiple surfaces can be created using manual dragging by three handles. In Figure 7-45, handle h_1 is moved along the curve C_1 to the dragged point. Dragging handle h_3 controls the length of the surface. Finally, handle h_2 changes the pull direction. The pull direction can be defined as one of the coordinate axes or by an arbitrary vector. Manual dragging can be replaced by definition of parameters. A flange and collar are created similarly. They start from a curve on the surface and extend at an angle to the surface normal and for a given distance.

7.3.5 Creating Surfaces by a Curve Network

Complex surfaces are modeled with specified continuity maintained by the model creating procedure across boundaries of the surfaces in the network. A surface is built up by individual surfaces. Construction curves for surfaces are generated by using segments of network curves. *Closed regions* are fitted with individual surfaces. Four-sided regions are enclosed by a boundary

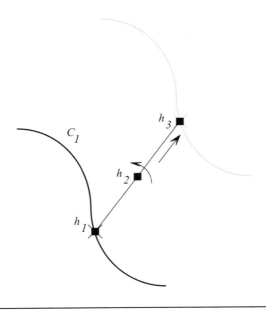

Figure 7-45 Creating a surface starting from a curve in a direction.

composed of four edges. In the case of three-sided regions, a surface is generated then trimmed. T-junctions are collapsed into four- or three-sided regions with four-sided and trimmed surfaces, respectively.

Construction of a curve network starts from characteristic curves of the surfaces to be modeled. Curves must intersect within a tolerance and closed chains of curve segments must bound closed regions. Three intersected curves are necessary as a minimum for the creation of a surface. More than a single edge is not allowed between two vertices. Parameterization of the surfaces is often matched, especially before knot insertion or other changes of curve network parameterization. Curves in a curve network are constructors for the surfaces. Curves can be removed from the network and new curves can be added to it to *change the surface design* or *create surface variants*. Mesh curves can be edited only if they are not mapped to sculpt curves. The surfaces are rebuilt according to the changed mesh curves.

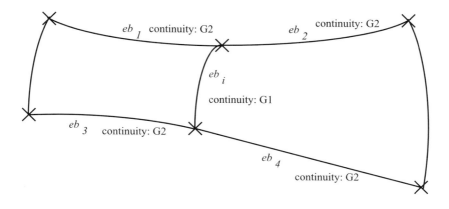

Figure 7-46 Continuity specification in a curve network.

Continuity is selected between surfaces across their common boundaries. This default continuity between the network based surfaces and with the surfaces from which the curves on surfaces, isoparametric curves, and boundary curves are extracted as curve network elements is enforced by the surface creation procedure. Moreover, local continuity can be defined across individual curves. Continuity at a boundary curve is dependent on other curvature definitions along the surfaces connected by it. In Figure 7-46, continuity G1 is specified for boundary edge eb_i. This continuity can be achieved only if the continuity across pairs of curves from the two surfaces (eb_1–eb_4) incident along this curve is at least G2. At the same time, this condition would be good for continuity G2 across eb_i.

The overall shape and character of the surface complex defined by network curves is tailored and modified by sculpt curves because some shapes are difficult to control by curves in a curve network. In Figure 7-47, two surfaces (S_1 and S_2) are created by using a curve network. Boundary edge eb_b bounds a single surface, while interior edge eb_i bounds two surfaces. The *effect of sculpt curve* C_s is usually defined by its level as large, medium, etc., instead of a numerical value.

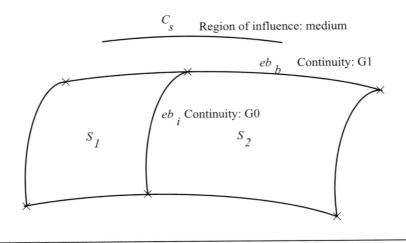

Figure 7-47   Surfaces and a sculpt curve.

A sculpt curve must be mapped to a network surface. Its points are projected onto the surfaces in a direction normal to them. The projected point is tied to the surface according to the parameter values of the point. The surface is modified by controlled modification of the sculpt curve. The parameterization of the sculpt curve is not allowed to change, consequently insertion and detachment of edit points and extension of the sculpt curve by another sculpt curve are also not allowed. These changes of the sculpt curve need its removal from the surface and then repeated mapping of the modified curve to it. However, any modifications of the shape of the sculpt curve by changing the position of the control vertices and edit points, constraining by aligning with any other curve, and smoothing are allowed.

The shapes of the surfaces within a network are modified by the shape of the actually mapped sculpt curve. At the same time, surfaces must interpolate the curve network. This interpolation can be restricted to the extent of modification by the sculpt curve. The sculpt curve can take control when the network surface is made free from one or more of its construction curves. For example, when the sculpt curve extends across an interior boundary edge,

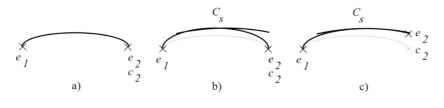

Figure 7-48 Effect of a sculpt curve.

then the original curve must be made ineffective for shape control. Edges that have been freed from the original curves are called *pinned edges*. Figure 7-48 shows a cross section of a surface between edges e_1 and e_2. The surface is controlled by sculpt curve C_s. The effect of the sculpt curve is limited if edge e_2 is connected to the original curve C_2 in the curve network (Figure 7-48b). Edge e_2 is freed from C_2 in Figure 7-48c.

Weight defines the influence of the sculpt curve. It controls the magnitude of the bond between points on the sculpt curve and their projections. The weight can be constant along the length of the sculpt curve or it can be specified at selected points. The weight profile is generated for the sculpt curve by interpolation between points of specified weights.

The effects of increasing the values of the parameters *weight* and *region of influence* are shown in Figure 7-49. The change of the shape is illustrated by the change of one of the border curves of a surface. The point P is on the sculpt curve.

Creation of a set of surfaces by a curve network is one of the most complicated surface modeling tasks. It is a suitable tool for solving most complex surface modeling problems, but some restrictions and numerous recommended considerations apply, depending on the task and the modeling system. Problems are generally associated with *triangular regions* and *T-junctions*. Curves for triangular sections are recommended to be simple without wildly varying shaped segments. T-junctions should be avoided. Curves across the network should be of the same degree and parameterization. The quality of the surfaces is better when continuity is homogeneous or less varied across the network.

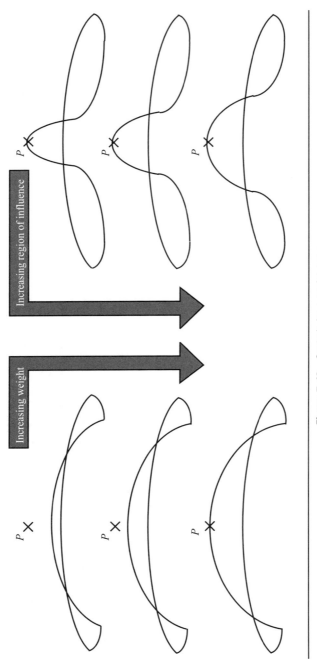

Figure 7-49 Controls by a sculpt curve.

7.3.6   Creating Free Form Surfaces Using Clouds of Points

Creating *free form surfaces from clouds of points* constitutes a bridge between the physical world and the modeled world. Physical objects are reverse-engineered by conversion of data from a *digitizer* and *measurement equipment* into a model *representation*. Point data in the physical world are acquired from laser or touch probe coordinate measuring machines and scanners. The measurements may give an *ordered mesh* or a *random cloud* of points. Point set information sometimes comes from experience based mathematical computation or existing geometry, or is imported in one of the standard data formats for point sets such as ASCII, G-Code, etc. Point set information is used for the creation of one or two *intersecting meshes of curves*. A mesh of curves can also be calculated directly, by using empirical formulas. At the same time, clouds of points can be created from existing curve and surface entities. Alternatively, a surface can be fit to a cloud of points.

In practice, all of the cloud points are used or only samples are applied. A surface can be created by the application of the *method of the maximum possible number of points*, by the *sampling of cloud points*, or by *starting from four corner points* of the surface. The method of the application of the maximum possible number of points starts with the definition of the boundary of the surface by the selection of an arbitrary number of curves. Continuity with connecting surfaces on boundaries can be kept by deformation of the created surface. A tolerance can be specified as the maximum distance between the surface and the cloud points together with the percentage of points that must be within the tolerance. The method of sampling is a simplified way of surface definition. A grid of sampling points is defined and applied for the creation of the surface. Typically, four boundary curves and several isoparametric curves are created from cloud points and then the surface is generated. The density of the grid and the number of spans can be specified in both of the parameter directions. Simple surface definition can be done by selection of four corner points for a four-sided surface. The surface is created within a specified tolerance, for a specified percentage of points.

Special modeling tools assist the construction, handling, visualization, and analysis of smaller or larger sets of points. Functions available for the processing of clouds include filtering, offsetting, extracting feature lines and curves, and removing extraneous points. Correlation can be defined between scattered data and the model. This allows for more effective processing of points. Sometimes an ordered set of points is required because certain modeling procedures cannot process random data. Large point clouds can be partitioned and sectioned into a few smaller clouds. Partitions are used for the creation and then blending of individual surfaces.

The resulting surface can be improved by *refined local re-digitization* of the physical object. Data sets can be improved by union and subtract combinations. In advanced modeling systems, dimensions can be defined as for dimension-driven free form surfaces within arbitrary points, in arbitrary directions.

7.3.7 Creating Models from Images

Most part models in industrial practice are constructed using planar contour based form features. Bezier or B-spline curves can be drawn on pictures and curves can be captured from contours found on pictures by many excellent software packages. It must be emphasized that this method has great potential for capturing shape ideas from arbitrary sources by simple computer tools for the modeling of products. Shape variants can be created and reviewed by engineers who have less modeling skill. This method can replace complex modeling for certain tasks. Sketches of shape ideas can be made by any simple computer drawing tool. Sketches and photos are converted into a geometric model by creating models of *protruded* and *depressed* surfaces from raster type images. Primary applications of this method are making logos, symbols, shapes, or reference numbers to be stamped or embossed on mechanical parts.

Tagged Image File Format (TIFF) is applied most frequently as the standard file format for this purpose because it saves embedded objects in the image. Objects are self-contained and identifiable components of the image, each with their own characteristics and attributes. They are applied individually to get information about the pictured or sketched outside world for the creation of models from TIFF images. The third dimension of shape can be controlled by settings such as the height of protrusions and depth of depressions.

7.3.8 Joining and Blending of Surfaces

In previous sections, several model construction tasks were discussed *where existing individual surface models* had to be composed into a *single surface model.* Now the main situations of joining and blending two existing surfaces are looked at in more depth (Figure 7-50). If a *common boundary* curve of the surfaces exists, the result can be a single surface or two surfaces according to the application. One or both of the surfaces are altered to achieve the specified continuity at the connection curve. If one of the surfaces is flat, cylindrical, spherical, or another analytic, alteration of its shape is not possible. If its representation is rational, the constraint *analytic* can be removed. Then the surface can be modified as a free form surface. Sometimes the common point must remain in its original position; this is an additional constraint to be considered at blending.

When a common boundary cannot be found, the surface can be connected to the other surface by *automatic snapping* to a nearby boundary. After matching their end curves by transforming one of them along a vector, the surfaces are handled as any other surfaces having a common boundary curve. If the surfaces to be joined are too far for snapping and transformation or modification of the surface for snapping is not allowed, a third surface is to be

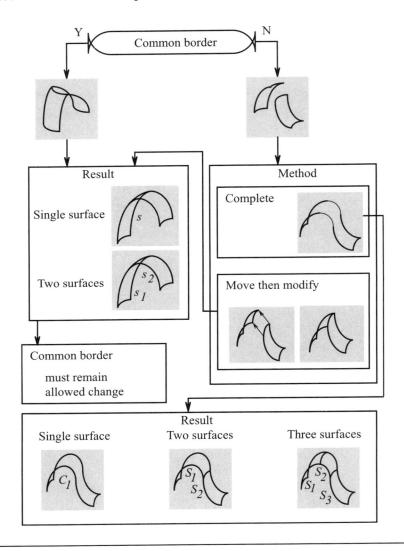

Figure 7-50 Joining and blending of surfaces.

defined to complete the complex surface. The result can be one, two, or three surfaces, according to the modeling task. The connecting surface is controlled during its creation to establish the specified continuities at both of the connection boundaries.

7.3.9   Modification of Surface Models

Well-defined *global surface modifications* recreate or replace surface model entities. This is especially important, where a fast *review of multiple design concepts* is necessary. Global modification of single or multiple surfaces acts directly on the entire surface or in one of the individually handled surface regions. Complex shapes composed of several surfaces, fillets, etc., can be modified with a single operation. Multiple surfaces can be globally changed as if they were a single surface so that time-consuming individual handling of surfaces can be eliminated. New elements can be connected to a complex surface. Modification of surfaces by change of *definition* or *value of dimensions* is one of the most important global alterations for engineering applications. Surfaces are trimmed by curves or extended according to curves, surfaces, and changed parameter ranges. They are split by curves or along their boundaries.

An effective high-level global model modification is *dynamic surface manipulation* where surfaces are dragged, pushed, twisted, and attracted by simple pointing device operations. A modified shape is achieved by interacting directly with surfaces instead of conventional modification by control points, knots, weights, etc. Significant changes can be done with little effort, even late in the design cycle. This is an economical way for the flexible design of products.

Fixed degree, maximum degree increase, or *optimized degree* can be the objectives at modification. Note that a modification that does not alter subdivision and degree of surface model will not increase the amount of model data. Non-uniform and nonlinear deformations of single and complex surfaces are considered problematic operations.

Most surface modifications are done by the modification of curves contextual with the modified single or multiple surfaces. Curve and surface changes are propagated onto other associative entities. Tedious, time-consuming, and troublesome manual prop-agation of complicated effects of surface modifications, from surface to surface, have been replaced by automatic, associativity driven propagation. Associativities as constraints between surface and curve entities, such as continuity and dimension specifications are saved by modification processes. This advancement has elimi-nated a source of serious and hidden errors and the need for extensive post-creation analysis of erroneous models. A changed continuity specification requires changes in curvature, slope, and inflections. Equations describing relationships between sur-face control variables can be applied to the associative control of shapes.

Surface models are modified by building them into the boundary representation of one or more solids. When a solid part model changes, surfaces associated with the part are changed accordingly.

7.3.10   Checking, Analysis, and Repair of Curves and Surfaces

Quality *improvements and evaluations* of surfaces demand the *checking and analysis* of surface characteristics. Sophisticated modeling procedures with built-in check and repair features have eliminated the extensive analyses needed after model creation pro-cedures. However, the results of these procedures and the proces-sing of imported models require post-creation analyses. The objective is analysis not only to reveal errors, but also to check the suitability of models and modeled objects for their application. Real time checking and analysis in advanced modeling procedures are based on the same principles as separate checking and analysis.

A great many problems are caused by imported geometry. The main sources of trouble in the development of the application of imported models are:

Incorrect or inappropriate models.
Inappropriate or erroneous exchange of model data.
Incorrect generation of an exchanged model in the receiving modeling system.
Human error in understanding the original model.

Values of various curve and surface characteristics are measured at given points, between given points and along a curve or a surface to facilitate checking and evaluation of curves and surfaces to decide their appropriateness for downstream applications. Advanced methods compare measured values with tolerance specifications. Tolerances are specified for the maximum allowed gap on and between curves and surfaces, for deviation of tangent and curvature values, and on either side of the minimum draft angle.

The results of measurements are applied for real time evaluation, by appropriate analyzing tools or by humans. Evaluations are followed by smoothing of surfaces, removing bulges and edges from surfaces, and other improving operations. Some improvements are done in real time with measurements and evaluations.

Dimensions, distances, and other characteristics are measured on curve and surface geometry. To do this, *measurement or location points* should be created on curves and surfaces according to the demand of the measurement. A measurement point can be placed in the model space or on an object, by a pointing device, according to x, y, and z coordinate data or u, v parameters of curves and surfaces. Frequently applied measurements are listed below.

Dimension and distance related measurements are:

Distance between two location points.
Angle between three location points.
Radius of curvature at any point of a curve.
Diameter of a circle defined on a surface.

Shortest distance between given points on two curves and surfaces.

Length of a curve or a curve segment.

How entities are fit related measurements are:

Gap between curves and surfaces.

Overlap or intersection of curves and surfaces.

Distance between curves and surfaces created from clouds of points and the points in the cloud.

Continuity related measurements within a surface and at the border of two surfaces are:

Gaussian and sectional curvature.

Surface curvature in different directions.

Position and size of gaps to be filled.

Positional, tangent, and curvature continuity deviations.

Application related measurements are:

Slope of a curve.

Area of closed contours and of single or multiple surfaces.

Surface normal at given u, v parameter values.

Draft angle in a pull direction.

Undercuts along surface, modified by draft form feature.

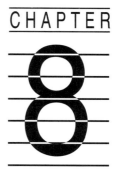

Construction and Relating Solid Part Models in CAD/CAM Systems

Part modeling is the historical area of computer aided design, drafting, and manufacturing. Traditionally, part models were constructed individually and each part model was stored in a *separate* file. Now, five different aspects *integrate* part modeling in a comprehensive product modeling environment. They are:

> *Associative modeling* of engineering objects.
> *Concurrent modeling* in *group work* of engineers.
> *Integration* of *wireframe*, *surface*, and *solid* modeling.
> Modeling *part variants and families*, starting from the same geometry.
> *Including knowledge* in part models.

During the 1970s and 1980s, computer aided engineering was concentrated on the modeling of parts and their manufacture. Complex surfaces and solids were represented by different methods: their models were often constructed in different modeling systems. While curves and surfaces were represented by early versions of Beziers and B-splines, solid modeling was based on an unevaluated form of constructive solid geometry with half space representations of solid primitives. Now, as explained above, unified model representation ensures integrated construction of parts and other objects of mechanical products. *Parts are modeled in an assembly context, surfaces are modeled in a part context, and kinematics are modeled in an assembly relationship context.*

In Figure 8-1, *Parts 1–3* are modeled. Part modeling is integrated by surface and assembly modeling. Contours C'_1 on *Part 1* and C_1 on *Part 2* are the same; they are created in the assembly space. Surface S_1 is created in a surface modeler and then integrated in *Part 2* as a form feature by substituting a flat surface with it. The surface S_1 has specified continuity for its boundary curves. Position continuity (D0) is specified at three curves, tangent (D1) continuity at the fourth one. Tangent continuity probably needs modification of the surface. *Part 3* is placed in relation to *Part 1* by three assembly relationships ar_1–ar_3. This is a typical group of relationships so the engineer selects it by *type of part placing,* then the modeling system identifies and generates the component relationships.

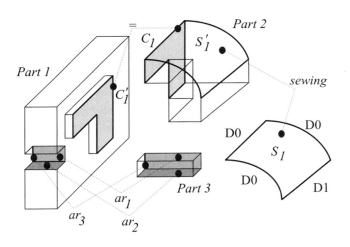

Figure 8-1 Integrated modeling.

When a *family of similar parts* is modeled, a common geometry should be shared by member parts whenever it is possible. The conventional method of mapping part instance data is a *part family table*. As a more advanced method, *reference part models* can be applied. Characteristics of parts are controlled with *design variables*. Form features are also created using design variables. Reference design variables are defined in reference models. Dimensions of a part instance are extracted from options stored for reference design variables. Parametric solid modeling can be combined with *inheritance* of design variables.

Integrated solid and surface modeling allows engineers to make shape representations in surface modelers. Closed complex skins can be created for solids and then they can be converted into solid models at any given point in the part modeling process. *Several parts* can be created referencing a *single surface model*. When a surface model is modified, all associated part models are modified accordingly. Moreover, some modeling systems allow *surfaces to be created directly on a part model*. This is the highest level of integration. Surfaces imported from other modeling systems can be sewn-up into solids directly, or after conversion.

Wire frames with unified topology and geometry use the same entities as solid models but without face and surface definitions. Note that these wire frames are not the same representations as the wire frames in the early era of geometric modeling. When a wire-frame model defines a shape unambiguously, a surface or solid model can be replaced by the simple wire frame for economical modeling. If necessary, a verified wire frame can be completed into a surface or solid.

Concurrent modeling in the form of *group work* of engineers is the advanced style of engineering. The assembly and its individual parts are developed concurrently, in an integrated process. Parts are modeled in an assembly context with user-controlled associativity. At the same time, part related specifications can be managed independently from assembly related specifications. The shape description of a part can be reused in other parts. Splitting a part into a multi-body supports both part and assembly design in teamwork.

Fundamental and useful *knowledge* from previous engineering practice can be captured to part models and utilized during improvements and applications of the model in the form of simple *rules* and *checks*. Rules are considered as specifications for specific features and parameters. Checks are defined as relationships within the model that can be verified automatically or on demand. They inform engineers of any violation and they validate the design. Results of checking can be applied at the generation of a proposal for correction procedures. Rules and checks may be defined before, during, or after the part design process. Equations, *IF-THEN-ELSE* rules, and other representations of rules and checks are captured, offered, and handled by advanced part modeling systems. They describe the steps of the design process, specifications of the product, earlier experiences of similar tasks, and design intent. Specifications may include standards, measures, etc.

Modeling tools in part modeling systems are included for the creation, integration, design analysis, model analysis, shape related modification, positional modification, application sequence modification (reordering), removal, and suppression of form features.

Form features are created as solids by a task based choice of volume modification, surface integration, and conditioning model construction operations as available in the modeling system. Hybrid modelers apply shape modification such as primary and volume combinations as an auxiliary means of part construction. Besides the integration of separately created surface, a surface can be constructed as a boundary of a solid by using one of the surface creation rules and as a swept, lofted, etc., volume feature.

Half of the *symmetrical parts* can be constructed and then mirrored using a selected flat surface sometimes called a mirror. The resultant half is associative with the original mirrored half. Any modification of the original half results in the same modification on the mirrored half.

Inertia and mass properties such as the absolute and relative inertial axis and center of gravity are calculated using geometry and density information. These calculations are based on the closed volume of a solid. Estimation of the cost for a part can be involved in part modeling by defining appropriate formulas and relating them with features.

8.1 Modeling by Form Features

Customer expandable choices of form features are available for part modeling in industrial modeling systems. Part modeling is a *sequence of shape modifications*. This process is based on a concept or modification steps are sequenced intuitively. Anyway, most of the sequence of modifications is governed by fundamental construction rules for each typical part. A form feature can be included in the model when its contextual elements such as existing form features to be modified or conditioned, as well as construction planes and other reference geometry, are available.

A *typical process of feature driven part modeling* is illustrated in Figure 8-2. Base feature FF_b is created as a tabulated solid, starting from contour C_{FFb}, sketched in its place in the reference plane RP_1. Feature FF_b is modified by volume adding features FF_p and FF_r starting from contours C_{FFp} and C_{FFr}, respectively. Rib feature FF_r needs reference plane RP_2 outside of the part. RP_2 is not included in the boundary of the part, but it is included in the part model. Features FF_p and FF_r are modified by draft conditioning features FF_{d2} and FF_{d1}, respectively.

Figure 8-2b shows two possible sequences of shape modifications. To complete the history of model creation, reference and construction entities are also included. The sequence can be changed by reordering during and after the design process of the part.

A *swept solid form feature* is a means of including a swept surface in the boundary of a solid (Figure 8-3). This operation is

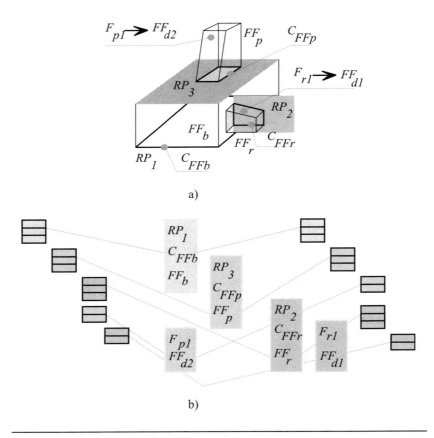

Figure 8-2   Part modeling by form features.

an alternative process to the creation of a swept surface in a surface modeler and then integration in the boundary of the solid as a form feature. Construction of solid sweep can start from open or closed generating and path curves.

A typical solid sweep is shown in Figure 8-4. A flange is created as a sweep feature along the inside boundary curve of a curved surface, in a boundary representation of a solid.

A *variable draft feature* on a flat surface changes its shape. In Figure 8-5, drafting creates a surface using neutral edge E_n as a spine. The surface is generated and then trimmed by its new boundary lines. The trimmed surface and new boundary curves

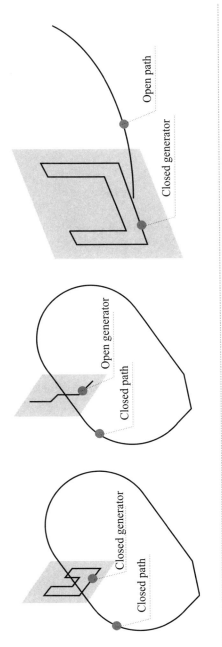

Figure 8-3 Open and closed contours for solid sweeping.

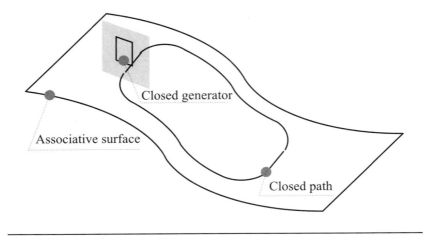

Figure 8-4 Sweeping along a surface.

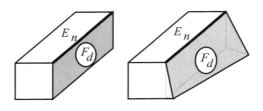

Figure 8-5 Serious effect of variable draft on geometry.

are mapped in the boundary representation to the appropriate topological edges and face.

Special considerations are needed for the sequence of shape modifications when *multiple conditioning is applied on a single volume feature*. Several general rules can be utilized such as:

The draft should be created before the fillet because continuity is to be kept at the connection of draft surfaces and fillets.

Shell and thickness features are often recommended to precede drafts and fillets.

A large radius fillet precedes a small radius fillet.

A constant radius fillet precedes a variable radius fillet.

All other fillets precede a transitional fillet.

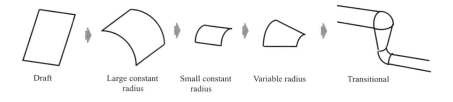

Draft Large constant radius Small constant radius Variable radius Transitional

Figure 8-6 Sequence of draft and fillet conditioning features.

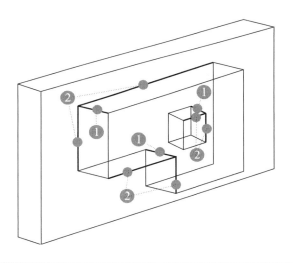

Figure 8-7 Sequence of filleting.

Figure 8-6 shows a proposed sequence of draft and fillet features. Where three filleted edges run into a corner on a solid, the sequence is recommended as illustrated by Figure 8-7.

Multiple application of the same form feature is often specified in the form of a pattern. Form features are placed according to a linear or rectangular *Cartesian grid* or a *circular grid* (Figure 8-8a). Form features FF_1, FF_2, FF_3, and FF_4 are multiplied along different grids. Modification of the master feature modifies the instances associatively. The number and data of the feature instances can also be modified. After changing the number of

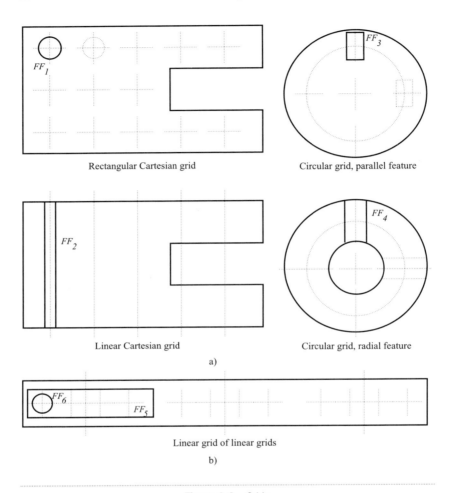

Rectangular Cartesian grid　　　　　Circular grid, parallel feature

Linear Cartesian grid　　　　　Circular grid, radial feature

a)

Linear grid of linear grids

b)

Figure 8-8 Grids.

instances, modeling procedures generally regenerate the grid. This operation simply multiplies the master feature and does not save any earlier modifications of feature instances. A new grid can be placed as an element of an existing grid (Figure 8-8b). Form feature FF_5 is defined as a linear grid consisting of four from the form feature FF_6. FF_5 is multiplied along a linear grid.

The procedure of part model construction is controlled by *associative modeling*. While a contour is being created for a form feature, its construction is supervised by a navigator. The navigator checks the required characteristics and parameter values. The workflow is traced by the modeling procedure; it allows going to the next step in the construction process only when entities created during the current step are acceptable. This method assures the quality of the model. Most form features are created starting with making a sketch in a construction plane. Construction of the contour goes ahead step-by-step according to selected design rules. The contour can be dimensioned and constrained on a single part or in an assembly context. The navigator offers alternatives for the next step of construction. In advanced sketchers, automatic constraint recognition and auto dimensioning assist the engineer. Constraints and parameters are possible to add after the feature creation. A constraint set can be refined by parameters and equations. At the same time, manual input of dimensions and constraints is also necessary. Although the modeler may allow over-constraining, one or more of the constraints must be removed in order to attain a correct and valued contour. Predefined single and multiple contours can be placed in libraries as frequently applied shapes for form features such as pockets, grooves, ribs, slots, stiffeners, etc.

Construction of a contour for a form feature in a part model and its validation by use of checks is explained in Figure 8-9. Contour C is sketched for form feature FF_1. Validation of contour C revealed three errors and communicated them to the engineer. Contour C contained a break point P_b. This is not allowed for the selected type of form feature; solid modeling requires a closed contour. Contour C was found to be open at point P_o, i.e., the gap is larger than the specified tolerance. Note that certain form features such as solid sweeps allow an open contour. Finally, the upper limit specified for the length L was exceeded. After correction, the modified contour was applied at the creation of form feature FF_1. Construction of the part model can be continued by form feature FF_2 because form feature FF_1 was proved correct during its repeated validation.

Modification of a part model is associative with assembly modeling, analysis, and the manufacturing process model.

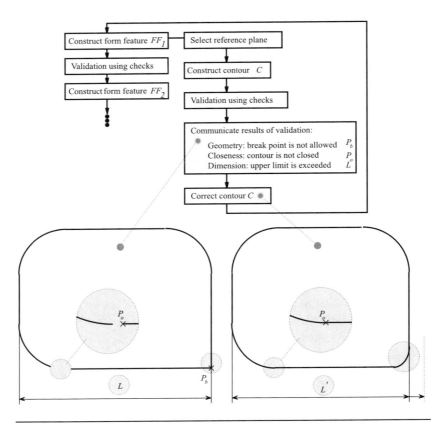

Figure 8-9 Part construction workflow.

The modeling environment where a given change of a part model causes changes, is called the *effect zone* of that modification. On the other hand, a feature can be *isolated* when its geometry is needed alone. In this case, the designer takes care that it is appropriately placed in its environment. There are several basic modifications associated with feature driven part models as follows.

Modification of the part. A *sequence of modifications* by form features can be reordered. Form features can be suppressed, deleted, or renamed. Modification of *dimensions on the level of a part* may affect the dimension and position of several features. In Figure 8-10a modification of L_2 modifies the dimension of

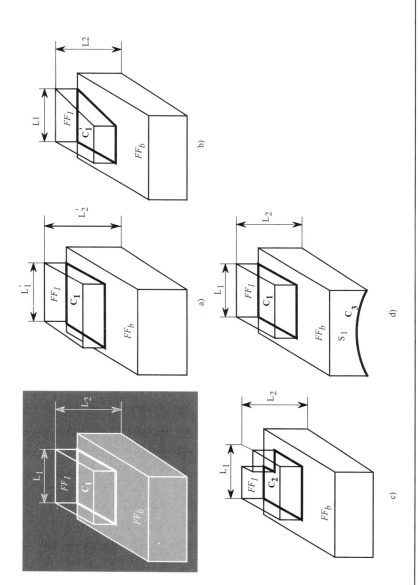

Figure 8-10   Modification of a feature based part model.

feature FF_b and position of feature FF_l. The constrained height of FF_l cannot be changed.

Modification of surfaces on a solid. When a designer modifies a curve or a surface, the system can automatically update the related solid (Figure 8-10d). Geometric entities in the support structure may also alter. In Figure 8-10d, replacement of a straight line by the planar curve C_3 changes a flat surface into a 3D surface S_l. When two or more part models are created referencing the same surface model, modification of the surface initiates modification of all related part models.

Modification of a feature. Construction elements and the sketch of a support plane for an existing feature can be replaced by new ones. In Figure 8-10c, the contour C_l is replaced by contour C_2. The shape (Figure 8-10b) and dimension (Figure 8-10a) of a construction contour of a feature can be changed.

8.2   Construction of Sheet Metal Part Models

Although parts with advanced surfaces have gained widespread application in the new styled world of industrial products, the advantages of complex sheet metal parts have saved their position in the construction of machines, vehicles, heating systems, computers, home appliances, and other products. Complex sheet metal parts are composed of very simple shape elements. They receive their shapes by very simple manufacturing processes such as bending, folding, and punching. At the same time, their manufacturing requires special processes in their design. Special modeling tools are available in CAD/CAM systems for this purpose. Advanced sheet metal parts are represented as solid models. This facilitates creating sheet metal shells from solids. The angle, radius, and bend allowance can be specified for each bend. Bend limits, including non-planar limits are defined as necessary. Dimensional and geometric constraints are included during the creation of contours for the walls.

Three fundamental ways for the construction of sheet metal part models are:

Construction of walls in model space, by sketch in place.
Creating walls by extrusion of each elementary line in a polyline.
Creating a part model in the model space by selection of bend lines on a flat pattern.

Contours for walls can be associatively extracted from existing solid parts. This allows for sheet metal part design in the context of the surrounding or enclosed assembly. Dimensional associativities between sheet metal and other solid part models allow easy and correct dimension-driven design changes without tedious manual re-dimensioning.

Construction of typical sheet metal part model entities in a model space is explained in Figure 8-11. Contours C_1 and C_2, vectors v_1 and v_2, and line L_1 are created in their final place then applied for the creation of walls for the sheet metal part. Walls starting from bend lines BL_2 and BL_3 are created by extrusion of lines BL_2 and L_1, along vectors v_1 and v_2, respectively.

The extrusion of open sections as a polyline (Figure 8-12) is a quick method of creating multi-paneled parts with bends. Bend parameters can be automatically inherited amongst bends. Vectors show the direction of extrusion (v_1) and material thickness (v_2). They can be inverted at and after creation of the model.

The method of creating parts by the definition of bend lines on a flat pattern (Figure 8-13) is advantageous when a complex flat shape is available for multiple walls or complex contours and the form features on the walls are too complicated to create it in the model space.

Sheet metal form features are placed on walls by positioning relationships such as holes, cutouts, slots, flanged punches, notches, depressions, channels, webs, corner and other stress reliefs, etc. They can be created in the flat pattern because it is associative to the folded state. Features and completed walls with all the form features can be copied or repositioned. In certain

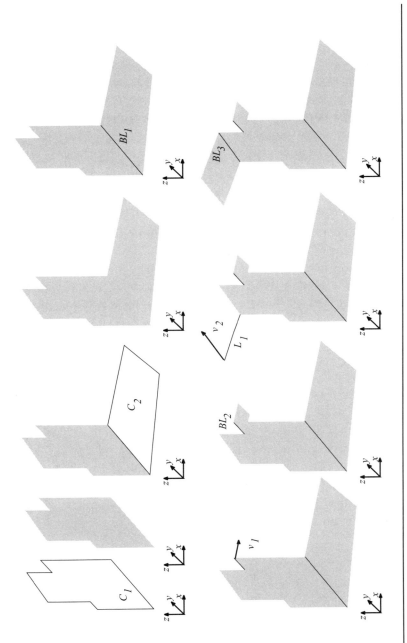

Figure 8-11   Construction of walls in model space by sketch in place.

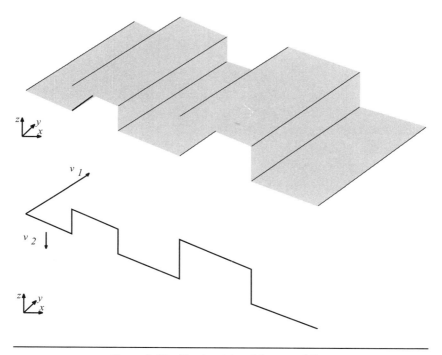

Figure 8-12   Sheet metal part from a polyline.

modeling systems, stress reliefs can be automatically calculated and placed by user-specified rules, etc. A set of standard relief shapes is stored as circular, rectangular, vee, etc. In Figure 8-14 contours C_1, C_2, and C_3 define sheet metal form features FF_1, FF_2, and FF_3, respectively, on a flat pattern.

Special construction features are represented as special information in the model. A few of them from industrial practice are as follows: zero radius bends allow for the modeling of bends that do not have an interior bend radius; zero degree bends represent edge connections where brazing or welding is specified instead of bending; 180° bends represent reinforcement bends.

The unfolded state of a part, or in other words the flat pattern, is fully associative with the part model. This is necessary because the part is cut together with punched features from a sheet metal table. Flattened parts are arranged on a sheet metal

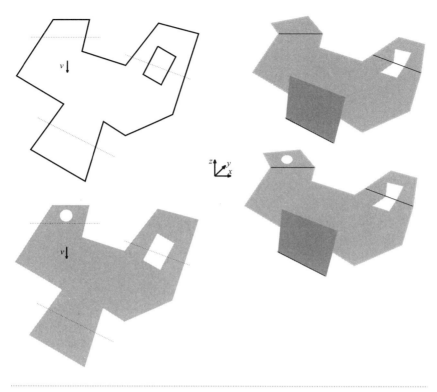

Figure 8-13 Sheet metal part from a flat pattern.

table so as to attain optimal material utilization. This process is called nesting. The change between the model space and the flat pattern representation is automatic. Flat patterns are annotated as production drawings, or applied for the calculation of tool paths for the numerical control of punching–bending press equipment.

Material growth for the flat pattern, or bend allowance or K-factor is calculated for the material, thickness, and machine information of each bend using generally applicable and user-specific bend tables and equations automatically.

Sheet metal part design is evaluated for fitting to its surrounding parts in assemblies, feasibility of manufacturing, interference with enclosed and covered parts, and overlaps in the flat pattern. Typical built-in and user-defined parts, walls, tabs, bends, and

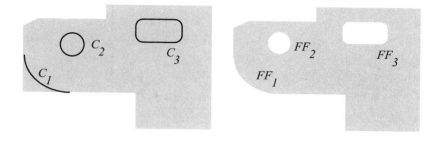

Figure 8-14   Sheet metal features.

form features are stored in libraries. Stored sheet metal parts can be applied for the creation of part variants or slightly different parts within a family of parts.

8.3   Creating Assembly Models

The main role of *assembly modeling* is the support design, modification, and analysis of complex assemblies. Conventional manual design and computer drawing techniques are not suitable for short product development cycles, including the fast modification and evaluation of designs. Assembly modeling activities in present advanced CAD/CAM practice include the following (Figure 8-15):

Creating the assembly structure.
Placing part instances in the assembly structure.
Defining attributes of part instances, such as part number, description, material, version, or revision.
Associating models to part instances.
Placing parts by assembly relationships, such as constraints between them.

These activities need an assembly modeling environment with an integration with part and other areas of modeling.

Advanced assembly modeling integrates part and assembly design in concurrent teamwork. This is the difference between

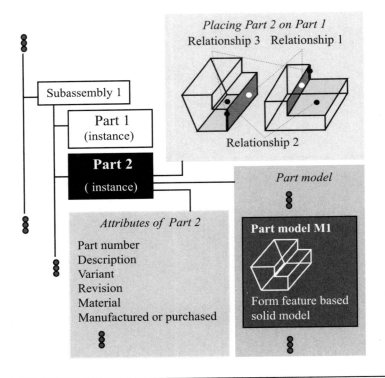

Figure 8-15 Assembly modeling activities.

the modeling of parts in CAD/CAM systems before and after the mid 1990s. Assembly modeling is an inherently teamwork activity because it integrates part modeling, kinematic modeling, the analysis of mechanisms, and production planning.

8.3.1 Assembly Structure

The *assembly structure* must be transparent for the engineer at any time during the creation, modification, and application of an assembly model. Advanced assembly modeling supports the easy survey and tracking of the whole assembly structure during assembly modeling and later by its *graphic visualization*. Structure descriptions often involve a very large quantity of parts according to the capacity of the modeling system.

Parts are modeled in the *context of assembly*. Associative part and assembly models help designers in understanding the product. At the same time, assembly structure description is independent of the representation of parts, and part models are managed independently from the assembly model. This is necessary to apply independent instances of a part model in different assembly models or several times in the same assembly model.

Theoretically, there are top-down and bottom-up approaches to the construction of assembly models. Construction by a top-down approach starts with the definition of the assembly structure followed by the creation of models of parts in the structure. On the other hand, bottom-up modeling assumes all part models to be available at the start of the assembly structure definition. *Conventional assembly modeling practice* follows a mixed approach where some parts are available at the start and are used at structure definition then additional parts are created for the modeled assembly. *Advanced modeling* also can be classified as a mixed approach. Form features of parts connected in the assembly are sketched in the assembly space. This may lead to the need for the construction of some new parts before placing any existing parts in the assembly. The structure is often constructed automatically on the basis of the placing of parts. Sketching assemblies in the model space without precise dimensions enables quick capturing of design ideas. Hand sketches of assemblies are processed by the same method as hand sketches of parts.

Functions for the creation and modification of assembly structures ensure navigation through huge model data structures to give quick movement across large assemblies. Assembly models are edited by *redefining, adding, removing, suppressing, replacing,* and *reordering of parts* in the structure. Cut, copy, paste, and drag and drop functions are also available in advanced systems. Some parts must be placed in two or more assemblies simultaneously, for example, when several variants of a product or families of similar products are modeled. To allow this, parts and subassemblies can be exported from an active assembly model to other assembly models. Similarly, functions are available for importing parts and subassemblies. The result is a flexible but well-organized and controllable simultaneous modeling in the form of group work. The access to part and assembly models is controlled by the group work manager through assigning engineers to projects and roles. Finished parts and assemblies are protected by changing their states to *under approval, approved,* etc., according to status definitions in the group.

Assembly modeling is supported by special-purpose visualization of parts, subassemblies, or a complete assembly while parts are being constructed and placed. Automatic generation of exploded views in connection with the bill of materials (BOM) serves the integration of assembly modeling with production planning. This connection between flexible product design and flexible product manufacturing assists in managing the short times allowed for putting new products into market.

8.3.2   Placing Parts by Assembly Relationships

Assembly relationships describe information about coincidence (Figure 8-16a), contact (Figure 8-16b), offset (Figure 8-16c), or angle (Figure 8-16d) relations between pairs of entities on placed and receiving parts. Related entities are lines or surfaces as selected on the shape descriptions or they are reference lines and planes. Where an appropriate entity is not available in the boundary of one or both of the parts, reference entities such as axes and planes are defined and used for the relative positioning of parts (Figure 8-16a). Any part in the assembly model should be fully placed (positioned, constrained) in relation to other parts. An assembly model with *under-constrained* or not fully placed parts is incomplete while *over-constrained* parts produce ambiguous assembly. Modeling systems continuously evaluate relationship definitions and generate messages for the engineer about under- or over-constraining.

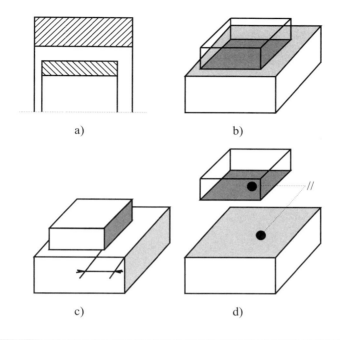

a)   b)

c)   d)

Figure 8-16   Assembly relationships.

Coincidence relationships define the coincidence of real or reference geometric model entities. In Figure 8-16a, two solids of revolution share the same axis. Contact can be defined at a point, along a line, or between two surfaces. The offset relationship gives information about distances specified between line, plane, or surface geometric entities. An angular relationship describes the angle or parallelism between geometric entities.

Figure 8-17 shows types of coincidence relationships between real and reference geometric model entities. They are concentricity (Figure 8-17a), point-on-line (Figure 8-17b), coaxiality (Figures 8-16a, 8-17c), and coplanarity (Figure 8-17d). Points coincide as centers of spheres SP_1 and SP_2 in the case of two concentric spheres. The center of sphere SP_1 lies on the line L_1. Ring shaped parts R_1 and R_2 have the same axis. Surface Pl_2 lies in the plane of Pl_1.

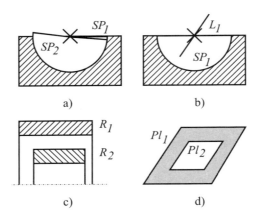

Figure 8-17 Coincidence relationships.

A contact relationship describes contact on a point, along a line, or on a surface (Figures 8-18 and 8-19). A common point, line, or surface is established between pairs of parts. Sphere SP_1 and surface S_1 are in contact at point P_1 (Figure 8-18a). Surface S_1 and cylinder C_1 (Figure 8-18b), cylinder C_1 and cylinder C_3 (Figure 8-18c) and cylinder C_1 and cylinder C_4 (Figure 8-18d) are in contact along lines L_1, L_2, and L_3, respectively.

Contact relationships are also defined between pairs of flat, cylindrical, conical, or spherical surfaces. Flat surfaces S_1 and S_2 (Figure 8-19a), cylinders C_1 and C_2 (Figure 8-19b), cones CO_1 and CO_2 (Figure 8-19c), and spheres SP_1 and SP_2 (Figure 8-19d) are in surface contact. Special contacts are defined along the teeth and threads of screws. Contact of bevel gears as in the case of CO_3 and CO_4 (Figure 8-19e) and CO_5 and CO_6 (Figure 8-19f) is defined as surface contact relationships.

Offset relationships describe distances between real or reference geometric elements. A line offset can be defined between axes (A_1 and A_2 in Figure 8-20a) or between an axis and a line (A_1 and L_1 in Figure 8-20b). Surface offset relationships describe offsets (O_1–O_3 in Figure 8-21) between a surface and a reference plane (S_1 and RP_1) or between two surfaces with the same (S_1 and S_2) or opposite (S_3 and S_4) orientation.

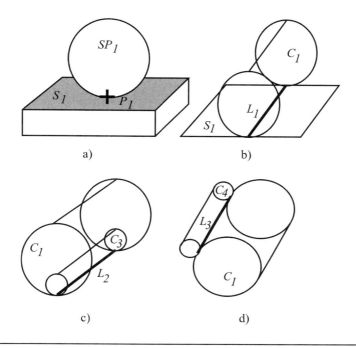

Figure 8-18   Point and line contact relationships.

An angle relationship defines an angle between the boundary and reference lines and surfaces. If the angle equals zero, the angular relationship is parallelism.

Besides constrained part placements, unconstrained part placements are also used for quick definition of assemblies by the manual placing of parts. Moving a part can be constrained by assembly constraints. Dynamic movement of parts allows their rotation, translation, snap into position, and drag and drop.

Figure 8-22 summarizes the application of assembly relationships for the placing of parts in an assembly by an example. Parts are placed using ten relationships. They are:

Part 2 is placed on *Part 1* by contact relationships numbers 1 and 2.

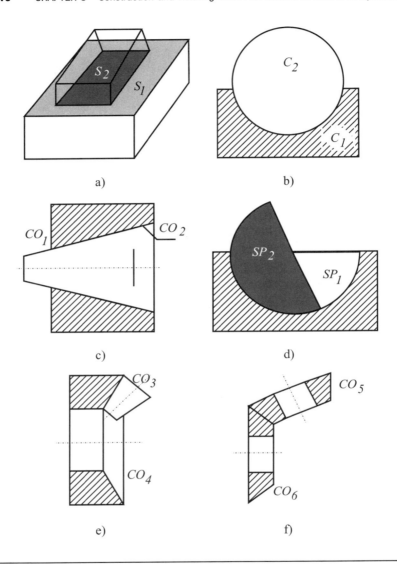

a)

b)

c)

d)

e)

f)

Figure 8-19 Surface contact relationships.

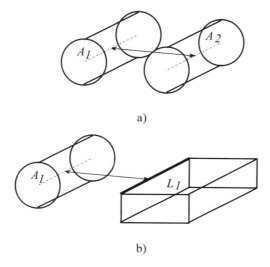

a)

b)

Figure 8-20 Line offset relationships.

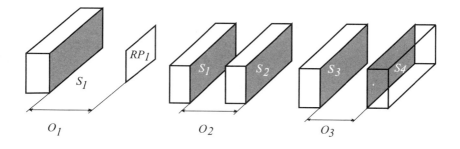

Figure 8-21 Surface offset relationships.

Part 3 is placed on *Part 2* by surface contact relationship number 3 and coincidence relationship number 4. Reference plane *Rp* is applied.

Part 4 is placed on *Part 2* by surface contact relationships numbers 9 and 10 and edge coincidence relationship number 8.

Part 5 is placed on *Part 1* by surface contact relationships numbers 5 and 6 and angle relationship number 7.

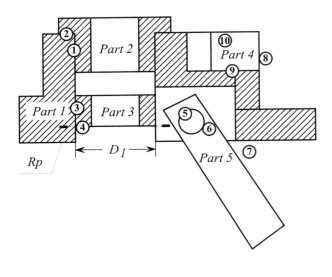

Figure 8-22 Application of relationships.

8.3.3 Reference Entities

Application of reference geometric entities in assembly models is illustrated in Figure 8-23. The relationship of *Part 1* and *Part 2* cannot be expressed using any boundary geometric elements. Axis *A*, an element that is not integrated in the part boundary, allows definition of a uniaxiality relationship in the model. Full placement is realized by offset relationship O_1. *Part 3* is positioned along axis *A* on *Part 1* using surface *S*. The same possibility is not available in the case of *Part 4*. Reference plane *Rp* should be included in the model and referred to in coincidence assembly relationship.

Similarly to the axial position, the relative rotational position of cylindrical parts is often impossible to define by boundary geometric entities. Figure 8-24 shows an example for a solution by the application of two reference planes, one is perpendicular to the axis of the cylindrical surfaces, and the other is in the face plane S_1 on *Part 2*. Three coincidence relationships are defined to place *Part 2*

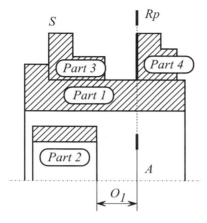

Figure 8-23   Real and reference elements.

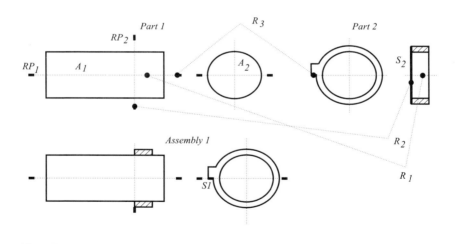

Figure 8-24   Application of reference planes on cylindrical parts.

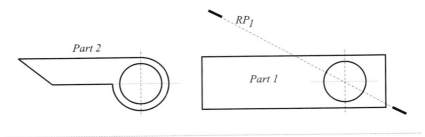

Figure 8-25   Application of a reference plane in the definition of an angular relationship.

on *Part 1*:

> R_1 is defined between axes A_1 and A_2 of the parts.
> R_2 is defined between reference plane RP_2 and surface S_2.
> R_3 is defined between reference plane RP_1 and surface S_1.

A typical application of the reference plane is the definition of angular relationships (Figure 8-25). Reference plane RP_1 controls the angular position at the placing of *Part 2* on *Part 1*.

The construction of assemblies is supported by numerous methods for solving a variety of problems that have emerged in engineering practice. Relationships are also defined for placing subassemblies. Form features that modify two or more parts, such as holes and channels going through several parts are called assembly features. Patterns, datum features, and surface copy features also can be defined as assembly features.

New parts can be constructed during assembly sessions by merging, cutting, or mirroring existing parts. Parts are grouped, moved, copied as instances, and mirrored. Construction can be controlled by an assembly family instance. Assembly families are created by using family tables. Simple tools, such as a *reference viewer* to see how components are referenced are also very important.

8.3.4   Analysis Using Assembly Models

Placing a new part in an assembly may cause inconsistency or over-constraining. Earlier defined assembly constraints may be broken

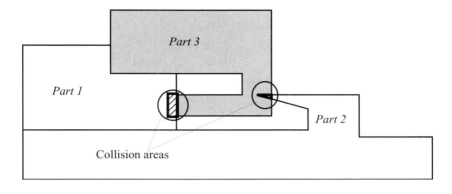

Figure 8-26   Collision areas at the placing of parts.

by design changes of parts. The consequences of these modifications are difficult to see. Fortunately, advanced methods are available in assembly modeling systems for the *analysis of assembly relationship networks*. Inconsistencies of the constraining are identified. Then attempts are made to reestablish broken contacts, automatically or by human interaction. Important analyses are those for checking specified distances and clearances between parts as well as detection of part-to-part collisions (Figure 8-26).

8.3.5   Part Modeling within the Context of Assembly

Creating a new part model often requires information from existing parts. The need for creating parts within the context of an assembly forced the development of assembly modeling systems to involve extensive part modeling in assembly modeling. New parts are created relative to existing parts. In creating a face on a new part, for example, the designer indicates that the new part is lying on the surface of an existing part. Contact relationships are created between the new and old parts. Then the contour for the new part can be extracted from existing parts. Development of part models during assembly modeling in the assembly model space is one of the most important recent contributions to the high level integration of part and assembly modeling. Part models are modified by considering the demands of their coexistence with other parts in an assembly. In the case of the multiple occurrence of a part in an assembly, any changes of the part model must be reflected in all its occurrences. This can be automated using model structure information. At the same time, minor variations between part instances can be handled.

Relationships are often defined between simplified part models. Then detailed modeling of parts is done in their own model spaces and detailed part models replace the simplified ones in the model database.

Creating Kinematic Models in CAD/CAM Systems

The model of the kinematics is the last subset of information required to complete the model of a mechanism. The first kinematics related decision is *place and type of joints*. The assembly model includes information for the places where the parts are in contact relationships. Joints are proposed to be placed at these connections automatically. This quick and easy construction of mechanisms has replaced the earlier separate definition of structures and joints. Part geometry and inertia properties are accessed from the assembly model.

9.1 Creating Joints

At construction of the kinematics, joints are defined between connecting geometries of parts. Although parts are considered as rod kinematic entities, their joining geometric elements are included in the construction of joint entities. A choice of practice oriented and user configurable joint[1] types is available in industrial modeling systems. They are associative with the geometry of connecting parts. The possibility of a clear display of existing joint entities is important to inform the engineer about types of joints, their place in the structure, degrees of freedom, and other parameters.

9.2 Analysis of Kinematic Models

Once joints, constraints, and functions have been defined, the mechanism can be analyzed automatically using a solver that understands and displays both kinematic and dynamic behaviors. *Consistency of motions* with the functional specifications, *degrees-of-freedom* calculations to evaluate the integrity of the design, solids based interference and clearance checking, the swept volume of a moving part which is defined by a part moving through its entire range of motion, and joint validity checking are the main automatic analyses.

User-defined kinematic laws allow time based simulation. Mechanism motions are sketched or defined using mathematical formulas. *Mathematical function manipulations* such as addition, subtraction, multiplication, division, scaling, integration, differentiation, and interpolation are important in the customization of calculations. Laws and relationships can be graphically visualized. Simulations analyze the speed and acceleration of coordinated movements of parts in different places of the mechanism as a reaction to specified input movement.

Repeated and combined simulations are important tools for the iterative development of mechanisms. Several motion, velocity, and acceleration functions can be displayed simultaneously for comparative analysis. Step-by-step, replayed, and recorded motion simulations are necessary for the analysis of results by engineers. At the position of a collision, the motion can be frozen for detailed analysis.

---

[1]Joint types were detailed in section 5.3.

Bibliography

Anand, V. B. "Computer Graphics and Geometric Modeling for Engineers," John Wiley & Sons, Inc., New York, 1993.

Anderl, R., and Trippner, D. "STEP. Standard for the Exchange of Product Model Data," Teubner, Stuttgart, 2000.

Au, C. K., and Yuen, M. M. F. A Feature Modeler for Sculptured Object Modeling, *Engineering with Computers*, vol. 19, no. 1 (2003): pp. 1–8.

Baartmans, B. G. "Introducton to 3-D Spatial Visualization," Prentice Hall, Englewood Cliffs, NJ, 1996.

Barequet, G., and Sharir, M. Partial Surface and Volume Matching in Three Dimensions, *IEEE Transactions on Pattern Analysis and Machine Intelligence*, vol. 19, no. 9 (1997): pp. 929–948.

Béchet, E., Cuilliere, J.-C., and Trochu, F. Generation of a Finite Element MESH from Stereolithography (STL) Files, *Computer-Aided Design*, vol. 34, no. 1 (2002): pp. 1–17.

Bloor, M. I. G., Wilson, M. J., and Hagen, H. The Smoothing Properties of Variational Schemes for Surface Design, *Computer Aided Geometric Design*, vol. 12 (1995): pp. 381–394.

Boehm, W., and Prautzsch, H. "Geometric Concepts for Geometric Design," A. K. Peters, Wellesley, MA, 1994.

Bronsvoort, W. F., and Jansen, F. W. Feature Modeling and Conversion – Key Concepts to Concurrent Engineering, *Computers in Industry*, vol. 21, no. 1 (1993): pp. 61–86.

Brown, K. N., Williams, J. H., and McMahon, C. A. Grammars of Features in Design, "Artificial Intelligence in Design," J. S. Gero (ed.), Kluwer Academic Publishers, Dordrecht, 1992: pp. 287–306.

Bu-Qing, S., and Ding-Yuan, L. "Computational Geometry: Curve and Surface Modeling," Academic Press, London, 1989.

Case, K., and Gao, J. Feature Technology: An Overview, *International Journal of Computer Integrated Manufacturing*, vol. 6, nos. 1–2 (1993): pp. 2–12.

Chang, T. C., Wysk, R. A., and Wang, H. P. "Computer-Aided Manufacturing," Prentice Hall, Englewood Cliffs, NJ, 1997.

Chen, C. P., and LeClair, S. R. Integration of Design and Manufacturing. Solving Setup Generation and Feature Sequencing Using an Unsupervised-Learning Approach, *Computer-Aided Design*, vol. 26, no. 1 (1994): pp. 59–75.

Chen, K.-Z., Feng, X.-A., and Lu, Q.-S. Intelligent Dimensioning for Mechanical Parts Based on Feature Extraction, *Computer-Aided Design*, vol. 33, no. 13 (2001): pp. 949–965.

Chuang, S. H. F., and Henderson, M. K. Using Subgraph Isomorphisms to Recognize and Decompose Boundary Representation Features, *Journal of Mechanical Design*, vol. 116 (1994): pp. 793–800.

Chung, J. C. Constraint-Based Variational Design, "Parametric and Variational Design," J. Hoschek and W. Dankwort (eds), Teubner, Stuttgart, 1994: pp. 63–70.

Ciarlet, P. G., and Lions, J. L. (eds). "Handbook of Numerical Analysis, Volume II: Finite Element Methods (Part 1)," North-Holland, Amsterdam, 2003.

Cohen, E., Riesenfeld, R. F., and Elber, G. "Geometric Modeling with Splines: An Introduction," A. K. Peters, Wellesley, MA, 2001.

Cook, R. D., Malkus, D. S., Plesha, M. E., and Witt, R. J. "Concepts and Applications of Finite Element Analysis," Wiley Text Books, New York, 2001.

Cutkowsky, M. R., and Tenenbaum, J. M. A Methodology and Computational Framework for Concurrent Product and Process Design, *Mechanisms and Machines Theory*, vol. 25, no. 3 (1990): pp. 365–381.

Danner, W. F. "Developing Application Protocols (APs) Using the Architecture and Methods of STEP (Standard for the Exchange of Product Data) Fundamentals of the STEP Methodology, US Department of Commerce, Technology Administration, National Institute of Standards and Technology, 1997.

De Martino, T., Falcidieno, B., Giannini, F., Hassinger, S., and Ovtcharova, J. Feature-based Modeling by Integrating Design and

Recognition Approaches, *Computer-Aided Design*, vol. 26, no. 8 (1994): pp. 646–653.

Deng, Y.-M., Lam, Y. C., Tor, S. B., and Britton, G. A. A CAD-CAE Integrated Injection Molding Design System, *Engineering with Computers*, vol. 18, no.1 (2002): pp. 80–92.

Dong, X., and Wozny, M. Instantiation of User Defined Features on a Geometric Model, "Product Modeling for Computer-Aided Design and Manufacturing," J. Turner, J. Pegna, and M. Wozny (eds), Elsevier Science Publishers, Amsterdam, 1991: pp. 183–195.

Dow, J. "A Unified Approach to the Finite Element Method and Error Analysis Procedures," Academic Press, London, 1999.

Falcidieno, B., and Giannini, F. Neutral Format Representation of Feature-Based Models in Multiple Viewpoints Context, "Product Modeling for Computer-Aided Design and Manufacturing," J. Turner, J. Pegna, and M. Wozny (eds), Elsevier Science Publishers, Amsterdam, 1991: pp. 165–182.

Farin, G. "NURBS for Curve and Surface Design," SIAM, Philadelphia, 1991.

Farin, G. "NURBS for Curve and Surface: From Projective Geometry to Practical Use," A. K. Peters, Wellesley, MA, 1994.

Farin, G., Hagen, H., Noltemeier, H., and Nödel, W. "Geometric Modeling," Springer-Verlag, Berlin, 1993.

Fischer, A. Multi-Level of Detail Models for Reverse Engineering in Remote, CAD Systems, *Engineering with Computers*, vol. 18, no. 1 (2002): pp. 50–58.

Floater, M. S. Parameterization and Smooth Approximation of Surface Triangulations, *Computer Aided Geometric Design*, vol. 14 (1997): pp. 231–250.

Fussell, B. K., Jerard, R. B., and Hemmett, J. G. Modeling of Cutting Geometry and Forces for 5-axis Sculptured Surface Machining, *Computer-Aided Design*, vol. 35 (2003): pp. 333–346.

Gallier, J. "Curves and Surfaces in Geometric Modeling: Theory and Algorithms," Morgan Kaufmann, London, 1999.

Gero, J. S. "Advances in Formal Design Methods for CAD," IFIP Book Series Volume 46, Kluwer, Dordrecht, 1996.

Greiner, G., and Hormann, K. Interpolating and Approximating Scattered 3D Data with Hierarchical Tensor Product B-splines, "Surface Fitting and Multiresolution Methods," A. Le Mehaute, C. Rabut, and L. L. Schumaker (eds), Vanderbilt University Press, Nashville, TN, (1997): pp. 163–172.

Groover, M. P. Automation, Production Systems, and Computer-Integrated Manufacturing, Prentice Hall, Englewood Cliffs, NJ, 2000.

Hagen, H., Heinz, S., and Nawotki, A. Variational Design with Boundary Conditions and Parameter Optimized Surface Fitting, "Geometric Modeling: Theory and Practice," W. Strasser, R. Klein, and R. Rau (eds), Springer, Berlin, 1997.

Henshaw, W. D. An Algorithm for Projecting Points onto a Patched CAD Model, *Engineering with Computers*, vol. 18, no. 3 (2002): pp. 265–273.

Hermann, T., Kovács, Z., and Várady, T. Special applications in Surface Fitting, "Geometric Modeling: Theory and Practice," W. Strasser, R. Klein, and R. Rau (eds), Springer, Berlin, 1997.

Hosaka, M. "Modeling of Curves and Surfaces in CAD/CAM," Springer, Berlin, 1992.

Hoschek, J., and Lasser, D. "Fundamentals of Computer Aided Geometric Design," A. K. Peters, Wellesley, MA, 1994.

Hu, S.-M., Li, Y.-F., Ju, T., and Zhu, X. Modifying the Shape of NURBS Surfaces with Geometric Constraints, *Computer-Aided Design*, vol. 33, no. 12 (2001): pp. 903–912.

Hughes, T. J. R. "The Finite Element Method: Linear Static and Dynamic Finite Element Analysis," Dover, New York, 2003.

Jablokow, A. G., Uicker, J. J., and Turcic, D. A. Topological and Geometric Consistency in Boundary Representations of Solid Models of Mechanical Components, *Transactions of the ASME*, vol. 115 (1993): pp. 762–769.

Jo, H. H., Parsaei, H. R., and Sullivan, W. G. "Principles of Concurrent Engineering, in Concurrent Engineering," Chapman & Hall, London, 1993: pp. 3–23.

Kief, H. B., and Waters, T. F. "Computer Numerical Control," McGraw-Hill, New York, 1992.

Kim, Y. S., and Wang, E. Recognition of Machining Features for Cast then Machined Parts, *Computer-Aided Design*, vol. 34, no. 1 (2002): pp. 71–87.

Kusiak, A., and Wang, J. Efficient Organizing of Design Activities, *International Journal of Production Research*, vol. 31, no. 4 (1993): pp. 753–769.

Kusiak, A., and Wang, J. Decomposition in Concurrent Design, "Concurrent Engineering. Automation, Tools, and Techniques," A. Kusiak (ed.), John Wiley & Sons, New York, 1993: pp. 481–507.

Laakko, T., and Mäntylä, M. Feature Modeling by Incremental Feature Recognition, *Computer-Aided Design*, vol. 25, no. 8 (1993): pp. 479–492.

Lauwers, B., Dejonghe, P., and Kruth, J. P. Optimal and Collision Free Tool Posture in Five-axis Machining Through the Tight Integration of Tool Path Generation and Machine Simulation, *Computer-Aided Design*, vol. 35 (2003): pp. 421–432.

Lee, K. "Principles of CAD/CAM/CAE," Prentice Hall, Englewood Cliffs, NJ, 1999.

Lee, K., and Lim, H. S. Efficient Solid Modeling via Sheet Modeling, *Computer Aided Design*, vol. 27, no. 4 (1995): pp. 255–262.

Lentz, D. H., and Sowerby, R. Feature Extraction of Concave and Convex Regions and Their Intersections, *Computer-Aided Design*, vol. 25, no. 7 (1993): pp. 421–437.

Li, C. L., and Hui, K. C. Feature recognition by template matching, *Computers and Graphics*, vol. 24, no. 4 (2000): pp. 569–582.

Liu, G. H., Wong, Y. F., Zhang, Y. S., and Loh, H. T. Error-based Segmentation of Cloud Data for Direct Rapid Prototyping, *Computer-Aided Design*, vol. 35 (2003): pp. 633–645.

Liu, T. H., and Fischer, G. W. Developing Feature-based Manufacturing Applicatons Using PDES/STEP, *Concurrent Engineering: Research and Application*, vol. 1, no. 1 (1993): pp. 39–50.

Männistö, T., Peltonen, H., Martio, A., and Sulonen, R. Modelling Generic Product Structures in STEP, *Computer-Aided Design*, vol. 30, no. 14 (1998): pp. 1111–1118.

Marcum, D. L. Efficient Generation of High-Quality Unstructured Surface and Volume Grids, *Engineering with Computers*, vol. 17, no. 3 (2001): pp. 211–233.

McMahon, C., and Browne, J. "CADCAM – From Principles to Practice," Addison Wesley, Reading, MA, 1993.

Mortenson, M. E. "Mathematics for Computer Graphics Applications: An Introduction to the Mathematics and Geometry of Cad/Cam, Geometric Modeling, Scientific Visualization, and Other Cg Applications," Industrial Press, Inc., New York, 1999.

Park, H. Choosing Nodes and Knots in Closed B-spline Curve Interpolation to Point Data, *Computer-Aided Design*, vol. 33, no. 13 (2001): pp. 967–974.

Piegl, L. On NURBS, A Survey, *IEEE Computer Graphics & Applications*, vol. 19, no. 1 (1991): pp. 55–71.

Piegl, L. "Fundamental Developments of Computer-Aided Geometric Modeling," Academic Press, London, 1993.

Piegl, L., and Tiller, W. "The NURBS Book," Springer, Berlin, 1997.

Pratt, M. J. Application of Feature Recognition in the Product Life-Cycle, *International Journal of Computer Integrated Manufacturing*, vol. 6, nos. 1–2 (1993): pp. 13–19.

Prautzsch, H., Boehm, W., and Paluszny, M. "Bezier and B-Spline Techniques," Springer, Berlin, 2002.

Rampersad, H. K. "Integrated and Simultaneous Design for Robotic Assembly," Wiley, Chichester, UK, 1994.

Ravi Kumar, G. V. V., Shastry, K. G., and Prakash, B. G. Computing Offsets of Trimmed NURBS Surfaces, *Computer-Aided Design*, vol. 35 (2003): pp. 411–420.

Roller, D. Foundation of Parametric Modeling, "Parametric and Variational Design," J. Hoschek and W. Dankwort (eds), Teubner, Stuttgart, 1994: pp. 63–70.

Roozenburg, N. F. M., and Eekels, J. "Product Design: Fundamentals and Methods," Wiley, Chichester, UK, 1995.

Rosen, D. W., Dixon, J. R., and Finger, S. Conversions of Feature-Based Design Representations Using Graph Grammar Parsing, *Journal of Mechanical Engineering*, vol. 116, no. 9 (1994): pp. 785–792.

Salomons, O., van Houten, F. J., and Kals, H. J. Review of Research in Feature-Based Design, *Journal of Manufacturing Systems*, vol. 12, no. 2 (1993): pp. 113–132.

Sandiford, D., and Hinduja, S. Construction of Feature Volumes Using Intersection of Adjacent Surfaces, *Computer-Aided Design*, vol. 33, no. 6 (2001): pp. 455–473.

Shah, J. J., and Mantyla, M. "Parametric and Feature-Based Cad/Cam: Concepts, Techniques, and Applications," John Wiley & Sons, New York, 1995.

Shah, J. J., Shen, Y., and Shirur, A. "Determination of Machining Volumes from Extensible Sets of Design Features in Advances in Feature Based Manufacturing," Elsevier Science Publishers, Amsterdam, 1994: pp. 129–157.

Sheffer, A. Model Simplification for Meshing Using Face Clustering, *Computer-Aided Design*, vol. 33, no. 13 (2001): pp. 925–934.

Shim, K. W., Monaghan, D. J., and Armstrong, C. G. Mixed, Dimensional Coupling in Finite Element Stress Analysis, *Engineering with Computers*, vol. 18, no. 3 (2002): pp. 241–252.

Soenen, R., and Olling, G. J. "Advanced CAD/CAM Systems. State-of-the-Art and Future Trends in Feature Technology," IFIP Book Series Volume 4, Kluwer, Dordrecht, 1994.

Taylor, D. L. "Computer Aided Design," Addson-Wesley, Reading, MA, 1992.

Toriya, H., and Chiyokura, H. "3D CAD Principles and Applications," Springer, Berlin, 1993.

Wang, C. C. L., Wang, Y., and Yuen, M. F. Feature-based 3D Non-manifold Freeform Object Construction, *Engineering with Computers*, vol. 19, nos. 2–3 (2003): pp. 174–190.

Wang, G.-P., and Sun, J.-G. Shape Control of Swept Surface with Profiles, *Computer-Aided Design*, vol. 33, no. 12 (2001): pp. 893–902.

Xue, D., and Dong, Z. Feature Modeling Incorporating Tolerance and Production Process for Concurrent Design, Concurrent Engineering: Research and Applications, vol. 1, no. 2 (1993): pp. 107–116.

Ye, X., Jackson, T. R., and Patrikalakis, N. M. Geometric Design of Functional Surfaces, *Computer-Aided Design*, vol. 28 (1996): pp. 741–752.

Ye, X., and Nowacki, H. Ensuring Compatibility of G^2-Continuous Surface Patches Around a Node Point, *Computer-Aided Design*, vol. 13 (1996): pp. 931–949.

Yu, X., Zhang, S., and Johnson, E. A Discrete Post-processing Method for Structural Optimization, *Engineering with Computers*, vol. 19, nos. 2–3 (2003): pp. 213–220.

Zeid, I. "CAD/CAM Theory and Practice," McGraw-Hill, New York, 1991.

Zha, X. F., and Du, H. A PDES/STEP-based Model and System for Concurrent Integrated Design and Assembly Planning, *Computer-Aided Design*, vol. 34 (2002): pp. 1087–1110.

Index